# ProGenesis
# Ninety-Five Theses Against Evolution

## A scientific critique of the naturalist philosophy

# ProGenesis
# Ninety-Five Theses Against Evolution

A scientific critique of the naturalist philosophy

by

Dr. iur. Dieter Aebi, Dr. med. Markus Bourquin, Prof. a.D. Dr.Ing. Werner Gitt, Roland Schwab, Dipl.Ing. Hansruedi Stutz, lic. theol. Marcel Wildi.

Copyright © 2013 Dr. iur. Dieter Aebi, Dr. med. Markus Bourquin, Prof. a.D. Dr.Ing. Werner Gitt, Roland Schwab, Dipl. Ing. Hansruedi Stutz, lic. theol. Marcel Wildi.

No part of this book may be reproduced or transmitted in any form or by any means, graphic, electronic, or mechanical, including photocopying, recording, taping, or by any information storage retrieval system, without the permission, in writing, of the publisher.

Strategic Book Publishing and Rights Co.
12620 FM 1960, Suite A4-507
Houston TX 77065
www.sbpra.com

ISBN: 978-1-61897-352-8

# TABLE OF CONTENTS

Preface .................................................................... xxi
Introduction .......................................................... xxvii

Biology (sixteen theses) .............................................. 1
  1  Micro and macro evolution ................................ 3
  2  Family trees and bushes .................................... 7
  3  Irreducible complex systems .............................. 9
  4  Mutation and the increase of information ........ 12
  5  Evolutionary mechanisms ................................ 15
  6  Biodiversity ...................................................... 18
  7  Symbiosis and altruistic behavior .................... 22
  8  Drosophila melanogaster ................................. 25
  9  Junk DNA ........................................................ 27
  10  Pseudo genes ................................................... 29
  11  Homeotic genes .............................................. 31
  12  Rudimentary organs ....................................... 33
  13  Recapitulation theory ..................................... 35
  14  Peppered moth ................................................ 37
  15  DDT resistant insects ..................................... 39
  16  Antibiotic resistance ....................................... 41

Geology and Palaeontology (fifteen theses) ...........................43
  17  Stasis in the fossil record................................................45
  18  Rapid fossilisation (taphonomy) ......................................48
  19  Missing Links...................................................................51
  20  Cambrian explosion.........................................................55
  21  Erosion of the continents.................................................58
  22  River deltas, sea coasts and reefs ...................................60
  23  Eruption of Mount St. Helens .........................................63
  24  Modern sedimentology....................................................65
  25  Undamaged layer boundaries..........................................68
  26  Polystrate fossils..............................................................71
  27  Living fossils...................................................................74
  28  Million year old artefacts ................................................76
  29  Million year old microbes ...............................................79
  30  Nusplingen platy limestone.............................................81
  31  Rapidly rising granite diapirs..........................................84

Chemical evolution (nine theses)..............................................87
  32  Vivum ex vivo .................................................................89
  33  The Miller Experiment ....................................................91
  34  Deoxyribonucleic acids (DNA)........................................93
  35  Polymer chemistry...........................................................95
  36  Chirality...........................................................................97
  37  Folding of proteins .......................................................100
  38  Addressing of proteins..................................................103
  39  Production of proteins ..................................................105
  40  Cell-internal control mechanisms.................................107

Radiometry and geophysics (eleven theses) ......................... 109
  41  Deviations in the radiometry ........................................ 111
  42  Accelerator Mass Spectrometer (AMS) ..................... 113
  43  Uranium, helium and lead in the zirconium ............... 115
  44  Radioactive decay to lead .............................................. 117
  45  Radioactive decay at plasma temperatures ................. 120
  46  Uranium and polonium radiohalos ............................... 122
  47  Helium from within the Earth ....................................... 125
  48  The Earth's magnetic field ............................................ 128
  49  Salt mountains and salt content of oceans .................. 131
  50  Nickel in seawater ......................................................... 133
  51  Fossil oil, coal and petrified wood ............................... 135

Cosmology and the big-bang theory (thirteen theses) ........... 139
  52  Singularity and Inflation ............................................... 141
  53  The formation of galaxies ............................................. 143
  54  The formation of stars ................................................... 145
  55  The formation of planets ............................................... 147
  56  surfaces of planets and moons ..................................... 149
  57  Precision planetary system ........................................... 151
  58  Earth to moon distance ................................................. 153
  59  Planetary rings .............................................................. 155
  60  Short period-comets ..................................................... 157
  61  Supernova remnants ..................................................... 160
  62  Metallicity of distant objects ........................................ 163
  63  The anthropic principle ................................................ 165
  64  Microwaves – background radiation ............................ 167

Philosophy (eleven theses) ..... 169
  65 Paradigm of evolution ..... 171
  66 Naturalistic world view ..... 173
  67 Dogma of the theory of evolution ..... 175
  68 Evolutionary psychology ..... 178
  69 Random processes ..... 180
  70 Causal evolutionary research ..... 184
  71 Homologous organs ..... 187
  72 Natural perfection ..... 189
  73 Teleology and orderliness ..... 191
  74 The meaning of life ..... 193
  75 Inappropriate beauty ..... 195

Information theory (eight theses) ..... 197
  76 Intelligent information ..... 205
  77 The Omniscient sender ..... 207
  78 The Powerful sender ..... 210
  79 The non-material sender ..... 212
  80 Rebuttal of materialism ..... 213
  81 Rebuttal of the big bang theory ..... 215
  82 Abiogenesis and macroevolution ..... 217
  83 Old and new proof of the existence of God ..... 220

Humans and culture (twelve theses) ..... 225
  84 Reports of the flood ..... 227
  85 The age of humanity ..... 232
  86 Neanderthals and australomorphs ..... 236
  87 The human and chimpanzee genomes ..... 238

88 Upright gait .................................................................242
89 The human eye ...........................................................246
90 The Inverse retina.......................................................248
91 The degeneration of human speech ...........................250
92 Human consciousness ................................................253
93 Human creativity ........................................................256
94 Conscience and ethics ................................................260
95 Love, joy, suffering and sorrow.................................263

Final declaration................................................................265
Epilogue .............................................................................267

# NINETY-FIVE ONE-SENTENCE THESES AGAINST EVOLUTION

1. The development of living organisms forming new species (macroevolution) by formation of new kinds of organs and structures has never been observed.
2. Research is revealing more and more unsystematically distributed characteristics of living organisms, so that the hypothesis of a genealogical tree of species is considered to be refuted.
3. No mechanism is known to explain the irreducibly complex systems that occur in living organisms.
4. Of 453,732 documented mutations described in approximately nineteen million scientific papers, only 186 were categorized as beneficial and none resulted in an increase of genetic information.
5. The known evolutionary mechanisms of mutation, selection, gene transfer, recombination of gene segments, gene duplication and other factors can not bring forth new body plans or functions.
6. Work sharing and mutual dependence, as observed in many plant and animal species within an ecosystem (biodiversity), cannot possibly have developed in small steps.
7. Symbioses and altruistic behavior of various plants and animals cannot be explained based on the known mechanisms of evolution.
8. Over three thousand artificial mutations in the fruit fly, Drosophila melanogaster, since 1908 have not produced a

single new, advantageous body plan; the fruit fly remains a fruit fly.

9. It is becoming increasingly evident that a great deal of the so-called junk DNA, designated until recently as "evolutionary garbage" as a result of evolution, does indeed perform specific functions.
10. More recent research is strongly suggesting that so-called pseudo genes, long considered non-functional, and therefore contradicting creation, do indeed have certain functions.
11. The hope that homeotic control genes would prove to be the key genes in macro evolutionary processes was up to now not fulfilled.
12. Rudimentary (half-finished or non-functional) organs are not useless remnants of upwards evolution. Most of these organs have a specific function, others are evidence of degeneration, and they could have been created in the present shape.
13. Although the drawings to the biogenetic law promulgated by Ernst Haeckel (1834-1919) were shown to be deceptive during Haeckel's lifetime, they are still to be found today in many schoolbooks!
14. Many schoolbooks describe the numerical changes observed in populations of the peppered moth as proof of evolution; in fact, it is not even microevolution.
15. All DDT-resistant insects have not developed by evolution, but are genetic variants that have always existed and have always been resistant to this insecticide.
16. The fact that bacteria can develop resistance to antibiotics is not an example of upward evolution, since the mutations that result generally involve a loss of genomic information.
17. The stasis (standstill) often observed in the fossil record shows that basic types have remained essentially unchanged throughout most of Earth's history, contradicting evolution.
18. For an organism to become a fossil, it must quickly be covered by sediments and cut off from air, since it will

otherwise rot or decay; therefore, fossils have been formed quickly and are no argument for high age.
19. Conclusive missing links between fish and amphibians, between amphibians and reptiles and between reptiles, birds and mammals have not been uncovered after 150 years of fossil research.
20. The so-called Cambrian Explosion (simultaneous appearance of most phyla in the Cambrian period) does not validate the theory that the living organisms share a common ancestor, but confirms creation.
21. In view of the fact that natural erosion over a period of ten million years would have worn Earth's continents down to sea level, there could not possibly be any fossil-bearing rock strata older than this.
22. Not a single river delta on the planet is more than several thousand years old, which sharply contradicts an Earth alleged lasting billions of years.
23. The eruption of Mount St. Helens in 1980 produced geological formations that correspond for the most part to those purportedly created in a process requiring many millions of years.
24. The characteristics of most of the sedimentary strata that are visible and accessible to researchers provide evidence of brief and intensive stratification processes.
25. The boundaries of the succession of beds in geological formations normally show very little or no surface erosion, bioturbation or soil formation, which require a fast formation of these strata.
26. Polystrate fossils, tree trunks and fossil animals that extend through more than one geological stratum challenge the theory of a slow, gradual development of these strata and make evolution impossible.
27. The existence of so-called living fossils shows no progress of evolution during millions of years and casts doubt on conventional interpretations of the fossil record.

28. Discovery of human artefacts in geological strata over two million years old call into question the reliability of the conventional timetable.
29. The viable microbes often found in old salt and coal deposits are barely not as much as 500 million years old.
30. Recent knowledge gained in the field of micro evolutionary speciation (sub speciation) demonstrates how species diversity, related to fossil marine animals, in the Nusplingen Limestone could have developed within a few decades.
31. More recent observations and calculations strongly suggest that the known granite diapirs developed as much as 100,000 times faster than had been previously assumed.
32. "Omne vivum ex vivo" (all life comes from life), Louis Pasteur's statement, has not been disproved to date.
33. Hundreds of so-called Miller experiments (primordial soup simulations) have been unable to explain or prove the accidental genesis of life.
34. Laboratory experiments have demonstrated that a chance origin of DNA under primordial soup conditions, without a supporting matrix as provided by a living cell, is extremely unlikely.
35. Since a hypothetical primordial soup would certainly have contained water, it is not known how long amino acid chains, let alone complete proteins, could have been formed.
36. Since only left-turning amino acids can be used to build living cells, the genesis of cells by chance is not known.
37. The correct folding of proteins is an information-controlled process that is extremely unlikely to occur by chance.
38. It is not known how a random process could generate the correct addressing of proteins in the cells.
39. The mechanism that starts and stops the production of proteins in each cell has to function properly from the very beginning.

40. The intracellular control mechanisms act counter to any trans-specific development due to the fact that life is basically geared towards maintenance of the existing proteins (stasis).
41. In view of the fact that the age results obtained through different radiometric methods show systematic differences for the same rock, there must be a source of systematic error inherent in either method of measurement and/or in the evaluation of the results.
42. Measurements by means of accelerator mass spectrometry (AMS) of carboniferous materials such as graphite, marble, anthracite and diamonds indicate an age of less than 90,000 years, despite an alleged age of many millions of years.
43. Rock strata claimed to be thousands of millions of years old contain zircons, the age of which, based on their helium content, is probably only four thousand to eight thousand years old.
44. Besides uranium-238, fifty-two other elements also decay into lead-206 (with a half-life of several microseconds to several thousand years), which is not taken into account in the calculations used for conventional radiometry.
45. The radiometric methods to determine the age of rocks deviate from non-radioactive methods by several orders of magnitude, which questions the radiometric methods.
46. The frequency of uranium and polonium radiohalos in the granites of the Palaeozoic / Mesozoic is evidence of one or more phases of temporarily accelerated radioactive decay.
47. Based on the heat radiating from the interior of the Earth, the amount of helium emerging from the inside of Earth accounts for only four percent of the amount expected if the Earth is 4.5 billion years old.
48. The Earth's magnetic field has reversed polarity several times in the past and has decreased to the half which indicates that the planet is fewer than ten thousand years old.
49. If the current processes of the uptake and release salt to and from the world's oceans would have lasted for 3.5 billion

years, the oceans should contain fifty-six times more salt than they actually do.
50. Calculations based on the amount of nickel transported annually by rivers into the world's oceans and the current nickel content of the oceans indicate that the processes at work today could have continued for a maximum of 300,000 years.
51. The claim that the formation of oil, coal and petrified wood requires long periods of time has been experimentally refuted.
52. In view of the fact that no mechanism has been identified leading out of the so-called "singularity," the concept of the big bang theory must be considered entirely speculative.
53. The origin of the galaxies cannot be explained within the framework of the big bang theory.
54. The origin of the stars has still not been explained, despite constant assurances by many cosmologists.
55. How planets could have originated from a disc of gas and dust is both unclear and highly controversial.
56. The highly different surfaces of planets and moons cast doubt on the theory that they have originated from a homogeneous cloud of gas and dust.
57. A solar system 4.5 billion years old is practically inconceivable in view of the fact that some planets fall into chaotic orbits after only ten million years.
58. The measured increase of the distance between Earth and moon is so big that the moon would have to be 3.5 times farther from Earth assuming an age of 4.5 billion years.
59. It is remarkable that all four gaseous planets have rings, since the maximum age of such a ring is only some ten thousand years.
60. Our solar system contains a much smaller number of short-period comets than expected in a planetary system billions of years old.

61. There are fewer supernova remnants in the Milky Way than one would expect after many billions of years.
62. The systematic difference in metallicity between distant and near objects that is expected in the big bang model is not observed.
63. The incredibly precise/fine adjustment of the various natural constants required to make life on Earth possible in the first place cannot be the result of a blind accident.
64. The non-uniformity of the cosmic microwave background radiation shows a cosmic north and south pole and a cosmic equator, which means that we could be near the centre of the universe, which contradicts the big bang theory.
65. Modern science conducts its research within the paradigm of evolution (macroevolution, primordial soup and the big bang theory), the basic tenets of which cannot be proven.
66. Since it is impossible to determine precisely the point at which natural phenomena stop and supernatural phenomena begin, it is not possible to explain the world on naturalistic terms only.
67. The origin of the theory of evolution is philosophical in nature (enlightenment, rationalism, naturalism) as well as a religious dogma with a scientific varnish.
68. Many of the conclusions of evolutionary psychology prove to be circular arguments, or are formulated in a very vague and undifferentiated way, so that they can be considered as merely plausible-sounding stories that can be neither be confirmed nor denied.
69. The rationale offered for macroevolution, with a combination of the factors accidental mutation and necessary selection, is without substance due to the element of chance involved, (i.e., it makes no logical assertions).
70. Causal evolutionary research cannot possibly explain an incalculable and unforeseeable development, which according to its own theory, is based on pure accident.

71. Similarities (homologous organs) do not prove a common descent; all they demonstrate is that the same basic principles have been applied in different organisms.
72. The observation that left to itself nature knows no incomplete ecosystems, and that most organisms contribute to the welfare of the entire ecosystem, is incompatible with the notion of accidental development.
73. The proposition that all the innumerable cosmic and biological structures have arisen by chance contradicts the obviously teleological and planned character of the entire natural world.
74. The theory of evolution cannot answer the question as to the ultimate meaning of life.
75. The purposeless beauty seen in nature cannot be explained by the naturalistic approach.
76. The code found in all forms of life allows only one conclusion: that there is an intelligent originator/sender of this information.
77. The concept according to which DNA molecules are encoded far exceeds the capacity of any human information technology; it cannot possibly have originated by chance from inanimate matter.
78. The knowledge necessary to program DNA molecules is not sufficient to create life because it would also require the ability to build all of the necessary biological machines.
79. Because meaningful information is essentially a non-material dimension, it cannot have been derived from a material dimension.
80. Human beings are capable of engendering meaningful information, which is of a non-materialistic nature and therefore cannot have originated from the materialistic part of our body.
81. The claim that the universe emerged from a singularity (scientific materialism) contradicts the non-material dimension of information.

82. Since all theories of chemical and biological evolution require that the information originate solely from matter and energy, we may conclude that all of these theories and concepts about abiogenesis are false.
83. Proofs of God's existence can be derived from the natural laws of information in the universe and from the prophetic information in the Bible.
84. Flood reports from ancient cultures on all five continents show that at least one gigantic deluge has happened on Earth.
85. The remains left to us by our ancestors (such as stone tools) provide evidence for at most several thousand years of human history.
86. No undisputed intermediate has yet been found of the hypothetical common ancestry of apes and humans.
87. At least 75 million "correct" mutations would have been necessary to make a modern human or a chimpanzee from a common ancestor – a highly improbable scenario.
88. The upright gait of humans requires simultaneous and coordinated changes of several characteristics in the skeleton and muscles, which clearly contradicts the notion of unguided chance development.
89. The twelve million nerve fibres which connect the human eye with the brain have to each lead to a specific location to generate a correct picture in the brain, which is not possible with the mechanisms of evolution theory.
90. New research confirms that the arrangement of the light-sensitive cells in the human eye represents an optimal design, refuting earlier claims to the contrary and pointing to a creator.
91. Studies of ancient languages have revealed that they were complex at first and grew simpler over time, which contradicts the notion of an upwards evolutionary development of human beings.
92. Studies of so-called near-death experiences strongly suggest that human consciousness exists in a non-material domain and can not be explained with evolution.

93. The human capacity for technical and artistic creativity indicates that the human spirit could not possibly have emerged from matter.
94. Conscience and ethics are hardly things that would have evolved in a graceless fight for survival that has been going on for millions of years.
95. It is impossible to reconcile the existence of the phenomenon of love with the ideas underlying the theory of evolution.

# PREFACE

On 31 December 2008, just in time for the beginning of Darwin Year 2009, a two-page article appeared in the newspaper *Die Zeit* under the headline "Thank you, Darwin!" A further four complete pages were dedicated to the theme of evolution. The thanks were directed towards a man who was born 200 years ago and whose groundbreaking book *On the Origin of Species* appeared 150 years ago. The philosopher, Immanuel Kant (1724–1804) had already proudly asserted, "Give me matter and I will construct a world." Fifty years later the French mathematician and astronomer Laplace (1749–1827) also boasted to Napoleon, "My theories have no need of the hypothesis 'God.'"

These and other fathers of scientific atheism were searching for an explanation for the source of life, in which God no longer figures. The apparently redemptive answer was provided by Darwin, who made it conceivable to explain the development of life "in a natural way."

## Is evolution a viable concept?

Just a quick glance into the realm of living things shows us consistently high-quality, targeted concepts: the sperm whale, a mammal, is equipped to be able to surface from a depth of three thousand metres without succumbing to the dreaded decompression sickness. A huge number of microscopically small bacteria in our intestinal tract have built-in electric motors, which can operate forwards or in reverse. In most cases, life depends on the full operation of the organs (e.g. the heart, liver, kidneys). Incomplete, still-developing organs are useless.

Anyone thinking here in the Darwinian sense must know that evolution has no perspective in terms of targeting a futuristic functioning organ. The evolutionary biologist, G. Osche, observed quite correctly, "Living organisms obviously cannot, during certain evolutionary phases, behave like a factory owner and temporarily shut down for renovations."

## Where does life come from?

In the face of all today's evolutionary hullabaloo, one asks oneself: where does life really come from? Evolutionary theory does not have the vaguest explanation of how the animate can develop from the inanimate. Stanley Miller (1930–2007), whose "primeval soup experiment" has been quoted in every biology book since the '60s, admitted forty years later that none of the present day hypotheses on the origin of life are convincing. He describes them all as "nonsense" or "chemical inventions." The microbiologist Louis Pasteur (1822–1895) recognised something very fundamental, "Life can only come from life."

## Why were the ninety-five theses of this book written?

Advocates of evolution consider their doctrine concerning the origin of life and the world to be a scientific theory. According to Karl Popper, an empirical theory must be falsifiable. That is to say, even the theory of evolution must, in principle, be disprovable. That is why the theses contained in this book were written.

The strongest line of reasoning in science is always given when the laws of nature can be applied in the sense that they exclude a process or procedure. The laws of nature know no exceptions. For this reason, a "perpetuum mobile – perpetual motion machine" (i.e., a machine that operates continuously without an energy supply) is a product of fantasy. We know today, what Darwin could not know, that an almost unimaginable amount of information exists within the cells of all living things and, moreover, in the most concentrated form known to us. The development of all organs is information driven; all processes in the living being function in an information-driven way and the production of all bodily substances (for instance, 50,000 proteins in the human body) is likewise information driven. The evolutionary thought system could only function at all if there was, in matter, the possibility of information being created by random processes. Information is not a property of matter:

Information is a non-material quantity; it is, therefore, not a property of matter. The natural laws of non-material quantities, especially those of information, signify that matter can never create a non-material quantity. Furthermore, it implies that information can only come into being via an originator equipped with intelligence and will. Thus, it is already clear that whoever considers evolution to be a viable concept believes in the "perpetual motion of information" (i.e., in something prescribed by the universally valid laws of nature). I will deal further with this in the chapter "Information Theory (theses 76–83)", which I have personally contributed to this book.

## Conclusion

The authors of the *Ninety-Five Theses against Evolution* have perceived that the teaching of evolution is one of the greatest fallacies in the history of the world. Were it merely a purely scientific question in some discipline or other, they would not have gone to such lengths to refute it. The reason is a different one. The question of our origin cannot be of no consequence to us, as it is intrinsically linked with the question of the existence of God. In terms of the authenticity of The Bible, only the two alternatives A1 and A2 are at issue:

A1: It is true that the origin and immeasurable diversity of life can be explained exclusively by the fundamental laws of chemistry and physics and the often-quoted evolutionary factors of mutation, recombination, selection, isolation, long periods of time, chance and necessity as well as death. Consequently, God is no longer necessary and The Bible would then be based on a completely godless source. It is a book conceived by human beings, and terms such as heaven and hell or resurrection and final judgement, derived from human imagination and have no relevance for human beings.

Or A2: What God told us in The Bible is true. Thus, the God of The Bible is the only living God and evolution a momentous scientific fallacy. Death is not a life-creating evolutionary factor, but rather a consequence of the separation from God (1). We can believe The Bible in its entirety, as Jesus prayed to God, the Father: "Your word is truth" (2), and as the Apostle Paul avowed, "I believe everything that stands written" (3). We will rise again after our physical death and have to answer before God; and there really is a heaven and also a hell.

## Purpose and aim of this book

The concept of the *Ninety-Five Theses* presented here is based unmistakably on the *Ninety-five Theses* of Martin Luther. At that time, he unleashed a revolution that had a worldwide impact. Luther stressed that The Bible came from a single divine source and, on the basis of this yardstick, was able to debunk the numerous injustices and false teachings of the Roman Catholic Church of the time. I hope these *Ninety-Five Theses* are equally effective.

Director and Professor (retired)<br>
Dr Werner Gitt

For almost 25 years until his retirement in 2002, Werner Gitt was Director and Professor at the Federal Physical-Technical Institute in Brunswick (Braunschweig, Germany).

## References:
1. Rom. 6:23.
2. John 17:17.
3. Acts 24:14.

# INTRODUCTION

Since the first publication of Charles Darwin's book *On the Evolution of Species*, on November 24, 1859, numerous facts have come to light that argue clearly against the theory of evolution. Yet the belief in evolution, the big bang theory and an Earth billions of years old has become deeply embedded in the consciousness of modern society. In the process, this ideology has gradually assumed a fundamentalist character. In no other area of science are critical voices attacked so personally and vehemently as in this field of research. Anyone expressing doubts is locked out of the debate on the origin of life and, not infrequently, vilified.

The intransigence among the upper echelons of science, education and the media is strongly reminiscent of the stubbornness with which the Roman Catholic Church of the Middle Ages defended its then worldview. On October 31, 1517, the reformer Martin Luther published his *Ninety-Five Theses* with which he challenged the practice of selling indulgences, widespread in his time. This intervention started a chain reaction, which finally led to The Reformation. In a similar way, the ninety-five theses presented here are intended to contribute to a rethink in the debate on the origin of life.

We wish, by means of this publication, to do all we can to ensure that in the debate on the origin of humanity (life on Earth and the universe), the free use of scientific data, interpretations and philosophical points of view * is made possible.

"If it could be demonstrated that any complex organ existed, which could not possibly have been formed by numerous, successive, slight modifications, my theory would absolutely break down." (Charles Darwin)

"There is no idea so absurd, that people will not quickly claim it as their own as soon as one has managed to persuade them that it is generally accepted." (Arthur Schopenhauer)

"In fact, men have only two questions to be answered: How did it begin and how will it end?"
(Stephen Hawking)

"The first drink from the cup of natural science brings atheism, but at the bottom of the cup waits God." (Werner Heisenberg)

\* From antiquity to the present, there is a vast and varied range of philosophical, ideological and scientific literature on questions concerning life and the universe. If you should discover within it a compelling disproof of one or more of the criticisms of the theory of evolution listed here, we request that you pass it on to us.

# BIOLOGY

The theory of evolution, as is taught in most schools today, states that all living creatures on Earth are related to one another and are supposed to have descended from single cell organisms and their precursors. Is this really true? What scientific evidence is there for this assumption?

Developments and genetic changes do actually occur in living creatures in the individual and in successive generations. In order to avoid misunderstandings, it is, however, necessary to draw a distinction between micro and macro evolution.

In micro evolution living organisms, during their biological history (as a species and an individual) already present structures and functions alter without the organisms basic blueprint changing in the process. In this way, over many generations, the wolf can evolve into a dog, and the famous Darwin finch can change the shape and size of its beak. Such modifications, however, always occur within a certain bandwidth, which cannot be exceeded.

In macro evolution, living organisms, completely new complex organs and functions not previously present could come into existence as a result of various occurrences within their genetic composition. In this way, in the past (over many generations and numerous intermediate stages), a simple single cell organism is supposed to have developed into a fish, then into a reptile, a bird, a hare etc. That such macro evolutionary processes have actually taken place*, must, after 150 years of evolutionary research, be seriously called into question.

* Successful new designs would have to integrate into the existing models of spatial, chronological and hierarchical genetic activity and would not be allowed, by intermediate stages, to disturb vitally important physiological, social, reproductive and ecological life patterns.

# 1

# MICRO AND MACRO EVOLUTION

There is not a single verifiable example of macro evolution. A succession of micro evolutionary events does not result in macro evolution because no new organs, structures or functions are created and no increase of information in the respective creature's genetic material takes place. Added to that is the fact that some micro evolution observed today proceeds 10,000 to ten million times faster than that generally derived from fossils.

Developments, genetic alterations in individuals and in the succession of generations really do occur. A good example of this is the development of breeds within a biological species. In this way, from the grey wolf species (Canis lupus) hundreds of breeds of dog originated, from Pekingese to St Bernard. Dog, however, remains dog. That is micro evolution.

Great variability is observed within a non-exceeded bandwidth. A broad development of breeds has taken place, especially with regard to domestic animals. With these, man has garnered mutations, crossed them with each other and selected according to his preference. In this way, for instance, the rock dove breed (Columba livia), in which Darwin took a lot of interest, has been subdivided into over a thousand breeds by breeders.

During his research expedition to the Galapagos Islands, Darwin collected different finch specimens. A remarkably large number of species of finch can be found on these Pacific islands.

In total, thirteen species, which differ significantly in body size, and beak size and shape, can be identified.. These Darwin finches are often invoked as proof of evolution in general, although it is indisputable that the changes in these birds lie within the domain of micro evolution. These different variations are still finches and will continue to remain finches (1).

## Macro evolutionary developments

According to Darwin's teaching on the origin of species, living organisms should, in their biological history, be able to create completely new, not previously present, complex blueprints, organs and functions by spontaneous events in their genetic machinery, mutation, selection, gene transfer, combination of gene segments, gene duplication and other factors. In this sense, the term "higher development" is generally used.

In an assumed case of higher development for instance; mammals from reptiles, structures such as hairs, milk glands, mechanisms for controlling temperature and everything else that differentiates mammals from reptiles would have to emerge.

These supposedly newly-created structures do not differ from the "old" structures by only one gene but usually by many genes, whose spatial and chronological activity patterns must be compatible with one another. With macro evolutionary processes, this would have to be the case with each individual intermediate form. The orchestra of genes must always be in tune.

## What is evolution?

The Dutch Standaard Encyclopedie under the heading "Evolution" reads as follows: "Macro evolution, which embraces the periodic appearance and dispersal of new groups as happens in the course of the geological periods, and which affects the higher orders of classification such as genera, families and orders or classes, cannot be accounted for directly by experimental genetics. The assumptive hypotheses based on the drastic alteration of genetic traits, cannot, in fact, be proved."

The causes, which are supposed over the course of millions of years to have led to the increased complexity of living creatures, are unknown. The biologist Willem J. Ouweneel concludes that genetics does not provide a basis for belief in macro evolution. In his opinion it emphasises quite the opposite: that the original forms of life, mostly the species but sometimes the genera and even perhaps the families, as variable as they may be, are as a whole constant and mutually discontinuous (2).

## Rapid speciation and/or micro evolution

That micro evolution takes place is not disputed. Indeed, it is also documented that it can proceed 10,000 to ten million times faster than that claimed for many fossil sequences (3) (4).

Contrary to previous assumptions, animals can adapt to changing environmental conditions within a few generations (5). Thus, it is clear that for palaeontologically proven micro evolutionary processes, from a biological point of view, no greater timescales were required (6).

## References

1. Helmut Schneider, *Natura, Biologie für Gymnasien*, Band 2, Lehrerband, Part B, 7-10. Schuljahr, Ernst Klett Verlag, (2006): 274.
2. Willem J. Ouweneel, *Evolution in der Zeitenwende*, Christliche Schriftenverbreitung Hückeswagen.
3. Virginia Morell. "Predator-free guppies take an evolutionary leap forward." *Science* 275, (28 March 1997): 1880.
4. Stephen Jay Gould. "The paradox of the visibly irrelevant." *The Lying Stones of Marrakech*, Frankfurt/M., (2003): 411–429.

5. Klaus Neuhaus, Schnelle Anpassung von Leguanen (Anolis) an neue Lebensräume (The rapid adaptation of iguanas to new habitats), Studium Integrale (1997/4): 81–83.
6. Uwe Brüggemann, Studium Integrale (1998/1): 38–39.

# 2

# FAMILY TREES AND BUSHES

Many traits of living creatures are allocated so unsystematically that, with increasing research, it becomes, not simpler, but more difficult to draw up consistent family trees and to reconstruct non-contradictory genealogical relationships. Instead of family trees, continuously new, stand-alone family bushes must be drafted. There is also the fact that modern DNA analyses force us to revise already accepted genealogical trees and to represent them once again as individual bushes. The creation of a generally acknowledged family tree of species has failed.

By drafting a single family tree of life (monophyletic representation), one is attempting to trace the origins of different forms of life (basic types) back to a single common ancestor. If, on the other hand, one talks of family bushes (polyphyletic representation), one means a plurality of individual lineages, which cannot be traced back to a single common ancestor (1).

In the past, one relied on drawing up family trees on the basis of anatomical and physiological features and characteristics of reproduction and behavior. Even then, it was often difficult to categorize the diversely distinctive genera, families and species of plants and animals in an unambiguous classification. Today, modern research also has the analysis of genetic makeup (DNA) at its disposal. Up until a few years ago, it was hoped that these DNA analyses would provide confirmation of the then-current family genealogical tree structures. These hopes were, however, quite unequivocally not fulfilled. The opposite was the case. Instead of the hoped for family tree, the designation of new family bushes came, ever increasingly, to the forefront.

## Family tree research in fossils

In spite of intensive research, to date not a single sequence of fossils has been found, which starts with the invertebrate and progresses via fishes, amphibians and reptiles to mammals (2).

## References

1. Reinhard Junker und Siegfried Scherer. *Evolution, ein kritisches Lehrbuch* (2006): 247.
2. Vij Sodera. *One Small Speck to Man, the Evolution Myth*. Vija Sodera Productions (2003): 37.

# 3

# IRREDUCIBLE COMPLEX SYSTEMS

An irreducible complex system is what one calls an arrangement of individual components, every one of which must be present for the overall system to function. For a car to run, it needs, at the very least, an engine, a clutch, four wheels, and a steering wheel. The idea that a primeval car in an early development stage, could have been driven without an engine or a clutch or without wheels is just as unthinkable as the idea that the biodiversity of life on Earth could have come about one step at a time.

All living creatures contain irreducible complex systems. If a single element of such a system is removed, either the whole system collapses or the entire function of the system comes to a standstill. Such systems cannot be set up one step at a time, as, without a certain minimum of components, they are incapable of functioning or living.

Charles Darwin has already addressed this problem. In his book *The Origin of Species* he writes, "If it could be demonstrated that any complex organ existed, which could not possibly have been formed by numerous, successive, slight modifications, my theory would absolutely break down."

In this respect, Darwin was still very level-headed. If he had known what we know today, then he would probably not have published his book. Indeed, in the last 150 years, the theory of evolution has become such a powerful myth, that many experts can no longer see the forest for the trees.

Even the simplest cell requires a special casing, mechanisms for controlling its metabolism, mechanisms for reading, writing and duplicating DNA, etc.

Further examples include human organs such as the eyes, the ear, or the brain; knee joints; the flying apparatus of birds, bats, and insects; numerous symbiotic life partnerships; the immune system; photosynthesis; intracellular protein transport, etc. (1).

The supply of individual components is also part of the production of these systems. They must be compatible with one another and be capable, from the very start, of performing their function correctly. The biochemist, Michael J. Behe states that, in the past seventeen years, not a single professional journal has reported anything concerning the intermediate forms necessary during the development of complex bimolecular structures (2). That should give pause for thought. Below are three examples of irreducible complex systems.

## The bacteria motor (3).

Certain bacteria are capable, with the assistance of a motor, of moving around. The rotor of this motor is connected to a whip, which can be made to turn, as a result of which the bacterium receives an impetus. This mechanism is made up of at least nine different individual parts, which must all be correctly assembled for it to function. That such a mechanism should develop one step at a time, with each individual intermediary stage bringing with it a practical survival benefit, is inconceivable. The numerous preliminary stages would not merely perform the function of movement less effectively, but rather not at all.

## Metamorphosis

The butterfly begins its life, just like flies, bees and beetles, in an egg out of which a caterpillar emerges. This caterpillar is preoccupied for the most part with eating. It grows rapidly and sheds its skin several times. Finally, an initial transformation from caterpillar into chrysalis takes place. In the chrysalis, a new creature comes into existence with completely new organs: the butterfly.

This transformation, known as metamorphosis, from caterpillar to chrysalis and then to butterfly, produces a irreducible complex system. Metamorphoses also occur in various different animal phyla, amphibians, coelenterates and others.

## Metamorphoses and change of host

Parasitic flatworms, such as the Lancet liver fluke, are known to have complicated transformation processes with several intermediate stages and host changes. Development cycles of this kind cannot have evolved via small development steps. It always needs all the chain links of development. If one is missing, the creature dies.

## References

1. Michael J. Behe. *Darwin's black box: The Biochemical Challenge to Evolution*. (New York: The Free Press, 1996) deutsche Übersetzung: Resch-Verlag, (2007): 87–225.
2. Michael J. Behe. *Nicht reduzierbare komplexe Systeme*. (factum Juli/August 1998): 32–39.
3. Michael J. Behe, *Darwin's Black Box*, (Resch-Verlag, 2007): 118–119.

# 4

# MUTATION AND THE INCREASE OF INFORMATION

According to the theory, macro evolution is supposed to have been driven by a random sequence of those mutations, which, in the organism's respective environment, proves to be an advantage to selection. In 2005, the biologist Gerald Bergman and his team searched through nearly nineteen million publications looking for beneficial mutations. Of 453,732 mutations described, only 186 could be classified as beneficial. Moreover, none of these mutations demonstrated an increase in genetic information for new functionally-capable proteins.

In conventional biology, there is a broad assumption that the number of different species that ever inhabited the Earth amounts to some $2 \times 10^{14}$ (200 billion). According to the advocates of evolution, for a new species to be created, an estimated thousand intermediate forms are necessary. Therefore, from an evolutionary theoretical point of view, to date, approximately $2 \times 10^{17}$ intermediate forms must have lived on Earth. To get from one intermediate form to the next would, once again, supposedly require a thousand beneficial mutations. That means that, by today, approaching $2 \times 10^{20}$ beneficial mutations must have been passed through.

That would be, calculated over the past 500 million years (during which evolution is supposed to have taken place) on average worldwide, 10,000 beneficial mutations per second! In spite of this, in the entirety of the specialist literature of the past decades, not a single mutation could be documented which would have added meaningful coding to the DNA (1) (2).

It must be taken into account that, in this representation, we are talking about successful mutations. According to the theory of evolution, a gigantic variety of random mutations would have to have taken place, for 10,000 successful ones to occur every second.

## Conclusion

It would be of vital importance to the theory of evolution for DNA strands to lengthen spontaneously and frequently. The fact that it has not been possible to establish the occurrence of such an event (with a beneficial effect for the creature) even after decades of sampling, may relate, among other things, to internal cellular-control mechanisms, which inhibit precisely this. Mutations can only survive this control process after duplicating if they are made up of the same number of building blocks as the original. Otherwise they will be instantly destroyed.

Richard Dawkins, a leading advocate of the theory of evolution, was asked whether he could give an example of the modification of an organism, in which information was added. He was not able to do so (3). Lee Spetner was of the opinion that "the inability to name even a single example of a mutation which added information signifies more than just lack of support for the theory" (4).

The fact is that even after over fifty years of intensive research, not a single example of the increase of intelligent information in the genome could be found.

## References

1. Gerald R. Bergman. "Darwinism and the Deterioration of the Genome." *CRSQ 42/2,* (September 2005): 110–112.

2. Barney Maddox. "Mutations: The Raw Material for Evolution?" *Acts and Facts* 36/9, (September 2007): 10–13.
3. Gillian Brown. "A Response to Barry Williams." *The Skeptic* 18/3 (September 1998)
4. Lee Spetner. *Not by Chance!* (The Judaica Press, 1997): 107, 131.

# 5

# EVOLUTIONARY MECHANISMS

The familiar evolutionary mechanisms of mutation (spasmodic alterations of the genotype), selection, gene transfer, combination of gene sections, gene duplication and other factors, do not suffice to explain the emergence of new blueprints and functions (macro evolution). These mechanisms are, practically without exception, ineffective or harmful, hardly ever useful and often fatal. In addition, there is the fact that, according to rough estimates by John Haldane, even one million years of constant development time would not be enough to produce a biodiversity such as we see today.

The mathematician, Lee Spetner, was able to demonstrate that the familiar and observed beneficial mutations (i.e., bacteria which build up resistance) always lead to a loss of information in the genome (1). In addition, Ronald Aylmer Sir Fisher has demonstrated that each individual mutation, even a beneficial one, can quickly be eradicated again by random effects (2). An individual mutation has a very slim chance of survival and would need about twelve million years to be integrated into the genome (3). The key question of causal research for evolutionary change remains, therefore, unanswered.

Darwin still believed in Jean Baptiste Lamarck's principle, according to which acquired traits are inheritable. However, as early as 1866 the Augustinian monk Gregor Mendel published a study in which he was able to prove that no new information comes into existence in the genome through inheritance but rather that information already in existence is merely newly combined (recombination). Today, Mendel's laws are undisputed.

## Haldane's Dilemma (4)

In the middle of the twentieth century, the famous evolutionist John Haldane attempted to employ so-called substitution load calculations. In doing so, he assumed that, through substitutions actual new basic types could come into existence. Then he attempted to calculate how much time that would take. He came to the result that even the most conservative estimates of one million years of constant development time would be nowhere near enough (5) (6).

However, one must bear in mind that the mathematical modelling of such population-genetic processes is extremely complex. Today, research is concentrating primarily on assessing the number of beneficial mutations that are actually identifiable. For more extensive calculations, important basic information is lacking to date.

## Spetner's approach

Mathematician Lee Spetner calculated the probability that a new basic type could come into existence as a result of random events in the course of macro evolution (7). On the basis of information from the established specialist literature, he came up with the inconceivable ratio of $1:3.6 \times 10^{2738}$. In comparison, it is estimated, that our universe contains approximately $10^{80}$ atoms. Thus, one would have to attach 2,600 noughts to the number of atoms in the universe in order to describe the probability ratio estimated by Spetner. Mathematician Emile Borel said that even at a probability of $1:10^{50}$, the event would be impossible..

Spetner is not alone in his ideas. Other scientists have obtained similar results (8). However, one has to be aware that, in this area of research, one is working with uncertain framework conditions and/or that due to the complexity of life, these are scarcely comprehensible. Such approaches can, however, give us an idea of the dimensions of the challenge.

## References

1. Lee Spetner. *Not by Chance!* (The Judaica Press, 1997): 20.
2. R.A. Fisher. *The Genetical Theory of Natural Selection.* (Oxford, 1958).
3. J.C. Sanford. *Genetic Entropy & the Mystery of the Genome.* (Elim Publishing, 2005) 126.
4. John B.S. Haldane. "The cost of natural selection." *Journal of Genetics* 55 (1957): 511–524.
5. Don Batten. "Haldane's Dilemma has not been solved." *Technical Journal* 19/1 (2005): 20–21.
6. G.C. Williams. *Natural Selection: Domains, Levels and Challenges.* (New York: Oxford University Press, 1992): 143–144.
7. Lee Spetner. *Not by Chance!* (The Judaica Press, 1997): 94–131.
8. G.L. Stebbins. *Processes of Organic Evolution.* (Englewood Cliffs: Prentice-Hall, 1966).

# 6

# BIODIVERSITY

The term biodiversity refers to the diversity of plant and animal species, the diversity within species and the diversity of ecosystems. Just as the human body is dependant on the division of labour between a multiplicity of cells and organs, an ecosystem, too, is dependant on the division of labour through biodiversity. For this reason, the scenario of gradual evolution, which is supposed to have started with a single cell, is unrealistic. It is conceivable that the ecosystems in which we live today must have been put together in a very short time, possibly within a few days.

In recent years, there has been a great deal of discussion and research in the area of biodiversity. The focus, in general, has been on saving and preserving ecosystems. That led to a wholly new understanding of and new methods of protecting endangered species. Instead of attempting to save individual species, one protects the entire ecosystems in which they appear. Thereby, species that are not so seriously threatened are protected at the same time.

The collective ecological service performed by the different species working for each other ensures that our planet remains clean and capable of supporting life. Biologist Yvonne Baskin writes, "It is this lavish array of organisms that we call 'biodiversity,' an intricately linked web of living things whose activities work in concert to make the Earth a uniquely habitable planet" (1).

It is impossible to draw up a full list of all ecological relationships. The most obvious ones are the food chains and the balance between oxygen and carbon dioxide maintained by plants and animals. Many degrading organisms make the soil fertile. Other biodiversity services purify the water, detoxify poisonous substances, regulate the climate, and pollinate the flowers.

Various experiments have been carried out in the course of research on biodiversity. In the course of these, it was discovered that highly diversified communities are more stable, more productive and resistant to stress (2) (3) (4). They have greater soil fertility and are, generally, in better condition.

## Redundant systems

An interesting phenomenon of ecosystems is redundancy, which means multiple back up of individual services. This means that a service performed by one species can also be undertaken by another. For this reason it was assumed that various redundancies make certain species superfluous (5). However, because all plants generally contribute both to soil fertility and to productivity, it is difficult to judge whether one can decide on the deficiency of one species on the basis of individual studies alone. What if this very species would also perform other services? In recent years, ecologists have turned away from talking about superfluous species; in fact, they even have a tendency to no longer use the word "redundant" (6).

With what we know today about biodiversity, it seems hardly possible that ecosystems or even life itself could exist without biodiversity and its eco-chemical and eco-physical services. The diverse services and the organisms that they offer must have existed side by side from the beginning because they form a complex system which cannot be reduced without penalty.

## Co-evolution as an explanation for ecology

As long as ecology appeared to be only a loose collection of organisms without connecting relationships, it was conceivable that it could have been built up by a gradual, directionless process. However, as more discoveries are made about the unbelievably complex network of biodiversity, the advocates of the theory of evolution see themselves as being in a similar dilemma as they were when the complex structure of cells was discovered. Because ecology is built upon such multi-layered, multi-species complexity, the explanation of its development, being a result of random events, represents a rather too painful challenge to our readiness to believe it.

In order to escape from this dilemma, the talk nowadays, when someone wishes to explain how ecology came about, is often of co-evolution. Co-evolution is defined as "the common evolution of two or more species," which "cannot interbreed and which have a close ecological relationship" (7). Regarding this, however, it must be noted that the ecological relationship precedes co-evolution. For that reason, co-evolution cannot be the answer to the question of the emergence of ecology. On this topic, biologist Henry Zuill writes, "I have no problem with two species fine-tuning an existing ecological relationship; I do have a problem with using co-evolution to explain the origin of ecological services. That is an altogether different problem. Remember, we are talking about an essential multispecies integrated service system—an entire integrated system. There seems to be no adequate evolutionary way to explain this. How could multiple organisms have once lived independently of services they now require?" Zuill continues, "It appears that life on Earth actually makes other life on Earth possible. That is, life on Earth makes it possible for life on Earth to proceed. This is not saying that life made (past tense) life on Earth exist, of course. It is saying that the whole system had to be present for life to go on existing. If this is true, there is no room for gradually unfolding ecology" (8).

# References

1. Yvonne Baskin. *The Work of Nature: How the Diversity of Life Sustains Us*. Washington DC: Island Press, (1997).
2. J.J. Ewel et al. "Tropical soil fertility changes under monoculture and successional communities of different structure." *Ecological Applications* 13 (1991): 289–302.
3. Shahid Naeem, Lindsey J. Thompson, Sharon P. Lawler, John H. Lawton und Richard M. Woodfin. "Declining biodiversity can alter the performance of ecosystems." *Nature* 368 (21 April 1994): 734–737.
4. David Tilman. "Biodiversity: Populations and Stability." *Ecology* 77 (1996): 350–363.
5. B.H. Walker. "Biodiversity and Ecological Redundancy" *Conservation Biology* (1992): 8–23.
6. Baskin, 20.
7. Robert Leo Smith. *Elements of Ecology*. (3. Auflage, Harper Collins): G-3.
8. Henry Zuill wrote in *Akte Genesis* by John F. Ashton, (1999) a contribution on the theme of biodiversity, which is the basis of this thesis.

# 7

# SYMBIOSIS AND ALTRUISTIC BEHAVIOUR

The familiar mechanisms of the theory of evolution break down when it comes to explaining the development of symbiosis and altruistic behaviour. Symbiosis is when both parties benefit from cooperation. Altruistic behaviour is when one of the parties serves the other and thereby even accepts disadvantage.

The greater part of the biomass on the Earth is made up of symbiotic systems. A large proportion of trees and bushes relies on pollination by other creatures, mostly insects. Then there are the lichens, a symbiotic life partnership between a fungus and green algae or cyanobacteria. Peter Raven from the Missouri Botanical Garden reports that when a plant dies out, ten to thirty others also become extinct (1).

## Some examples of symbiosis:

- The pollination of flowering plants by insects, whereby the insects receive nectar as nourishment.
- Ants protect aphids and receive sugar water in return.
- Mycorrhizal fungi take carbohydrates from trees and orchids and give minerals and water from Earth in return.
- The transportation of plant seeds by humans and animals, where the fruits are eaten and the seeds are discharged at a place distant from the plant.

- Many sessile, shallow water-living invertebrate sea creatures such as fire coral, most anemones and giant clams live with photosynthetic zooxanthellae.
- All mammals are absolutely dependant on gastrointestinal bacteria (e.g., escherichia coli).
- Lichens are a symbiotic structure made up of algae and fungi, where the algae produce carbohydrates by photosynthesis. These are absorbed by the fungi, while the fungi provide the algae with water and nutrient salts.

## Altruistic behaviour by the oak towards the gall wasp:

The gall wasp lays its eggs on the oak's leaves. The oak leaf then forms a small pod for the egg. The egg matures in this pod until finally a small caterpillar hatches. This caterpillar can then feed on the nutritious cells on the inside of the pod and is, at the same time, protected from birds. When it has grown large enough, it leaves its little house, pupates and becomes a wasp, which, in turn, will lay eggs on the leaves of the oak.

These small envelopes formed by the oak and other tree species are called galls. These galls are not only caused by gall wasps but also by other insects and mites. The creatures that cause the galls are not the only ones to benefit from them; parasitic fungi and bacteria do so as well.

The galls are nodule type growths, which are of no benefit to the plant itself. It is astonishing to what extent they are tailored to the necessities of the life of the animals and plants concerned. Yet various animals are able to trigger the formation of galls on the same plant, while the galls themselves develop different structures. This is not, therefore, in any way a general irritative effect as a reaction to the egg being laid, to a sting or the like (2).

The mechanisms of the theory of evolution would hardly support the development of galls, as they quite clearly entail disadvantages for the host plant. The complicated process of forming galls like this would, from an evolutionary theoretical perspective, more likely be selected out than advanced.

## References

1. Yvonne Baskin. *The Work of Nature: How the Diversity of Life Sustains Us*. (Washington DC: Island Press, 1997): 36–37.
2. Paul Lüth. *Der Mensch ist kein Zufall*. (DVA, 1981): 188–190.

# 8

# DROSOPHILA MELANOGASTER

The fruit fly, drosophila melanogaster, has been established as a model organism for genetics since 1908. Over 3,000 mutations of it have been documented to date. Yet, to date, not a single further development towards a new, beneficial blueprint has ever been established.

For one hundred years, biologists have been using the small fruit fly, drosophila, to carry out thousands of experiments to research the laws of inheritance (1). To this end, biology students have been working in their laboratories with fruit flies, where they attempt to produce new variants by crossing various fruit fly types.

Today, there are thousands of publications on fruit flies. It is the preferred organism for researching evolutionary genetics. It is used because it has a simple genetic structure and is easily bred in the laboratory. Additionally, it contains four pairs of easily studied chromosomes with only 13,000 genes. In March 2000, the complete sequence of the genome of the fruit fly was known (2).

## Artificially created mutations

Artificial mutations can be produced in the laboratory using x-rays. That was how, for instance, abnormal wing shapes and colored eyes were created. But despite countless mutations and intelligent human selection, not one new creature ever came into existence. Evolutionist Pierre-P Grassé concluded, "The fruit fly, the geneticist's favorite research object, the geographic, biotopic, urban and rural variants of which are known back and front, appears to have stayed the same since primeval times" (3).

## Adapting to a dry climate

Ary Hoffmann is Director of the Centre for Environmental Stress and Adaptation Research at La Trobe University in Melbourne (Australia). He wanted to know whether the Australian fruit fly drosophila birchii could adapt to a dry climate. In several experiments a group of flies was exposed to a very dry climate, so that ninety percent of them died. He bred the survivors further and exposed them once again to the arid conditions until, again, ninety percent died.

He repeated this for more than thirty generations. The expectation that these flies would adapt to a climate being much drier was not fulfilled (4). Hoffmann and his staff quite quickly identified the limit of adaptation of these flies. If the tropical climate in which these fruit flies lived were actually to become drier, it can be assumed that they would become extinct.

## References

1. Frank Sherwin. "Fruit Flies in the Face of Macro evolution." *Acts and Facts* 35/1 (January 2006): 5.
2. Mark D. Adams, et al. "The Genome Sequence of Drosophila melanogaster." *Science* 287, (24 March 2000): 2185–2195.
3. Pierre-P. Grassé. *Evolution of living Organisms*. (New York: Acad. Press, 1977): 130.
4. Terry Lane und Ary Hoffmann. In *Radio National, The International Interest*. http://www.abc.net.au/rn/talks/natint/stories/s911112.htm.

# 9

# JUNK DNA

Only a small part of DNA in humans, around five percent, is coded for proteins. Some years ago no function could be assigned to the remaining ninety-five percent, so it was over-hastily designated "junk DNA" (junk = rubbish). This junk DNA had been regarded as a confirmation of the theory of evolution; indeed, such evolutionary rubbish would be, as a by-product of a randomly driven evolution, expected. It has now been shown, however, that large parts of this junk DNA fulfil very well defined functions.

In recent years it has been debated whether junk DNA really is redundant. In the meantime, the findings are increasing, that non-coding sequences of DNA play an important role in the regulation of gene activity and cell division.

Clues have also been found pointing to the fact that non-coding DNA could play a role in the antiviral immunological strategy in the organism and, despite its simple structure, forms part of the immune system (1). In addition, it has been discovered that several sections of junk DNA already have, during early embryonic development, a large influence over the interaction of genes (2).

This is no isolated example. Several structures in living organisms were initially declared to be "pointless" or "rudimentary," but as a result of increasing research, their true ingenuity was discovered. Today, the generally predominant opinion among evolutionary researchers is that natural selection would long since have eradicated this apparently redundant DNA, if it had not fulfilled some task (3).

Today's known role of junk DNA rebuts the often-stated argument that life on Earth could not have originated from an intelligent creator.

## References

1. *John Woodmorappe. "The potential immunological function of pseudo genes and other "junk" DNA." Technical Journal 17/3 (2003): 102–108.*
2. Gill Bejerano. "Junk DNA Now Looks Like Powerful Regulator." *ScienceDaily* 24 (April 2007).
3. Markus Rammerstorfer. "Nur eine Illusion?" Tectum-Verlag (2006): 82.

# 10

# PSEUDO GENES

"Pseudo genes" (pseudo = lie) are structured like genes, look damaged and are mostly not used. For that reason, one saw in them the remnants of random evolution. With increasing research, it has, however, been demonstrated that several pseudo genes have important regulatory functions in genetic activity and embryonic development.

The discovery of a human retropseudogene *(a disused gene, which came into existence as the result of an unsuccessful copying process)* was a surprise (1). It is coded as an anti-tumour gene (an anti-cancer gene) that can recognise defence cells in the immune system (T cells).

Anti-tumour (tumour-suppressor) genes help the organism to destroy tumour cells. They demonstrate a varying degree of success. One important research objective is to increase their effectiveness in order to obtain therapeutic vaccines against cancer.

This realisation followed an earlier discovery of protein coding genes, which have the secondary capability of producing short segments of antigenic peptides.

Junk DNA and pseudo genes in the theory of evolution:

Today junk DNA and pseudo genes are fully integrated in evolution theory. However, the expressions "junk" and "pseudo" are not justified.

## Reference

1. *John Woodmorappe. "The potential immunological function of pseudo genes and other 'junk' DNA." Technical Journal 17/3 (2003): 102–108.*

# 11

# HOMEOTIC GENES

Homeotic genes are control genes that set in motion the entire developmental cascade of embryonic development. They are very similar over a wide systematic range, from the fly, mouse, chicken, right up to the human being. The great similarity of these embryo-developmental control genes initially led one to think of them as key genes of macro evolution. This expectation, however, remained unfulfilled.

Some decades ago, a biologist at the University of Denver, in the course of a public debate, proclaimed an example of a "beneficial" mutation. It concerned the bithorax gene that produces four wings in the fruit fly. This, however, reduces its capacity to fly. Possibly the control program in the brain for flying with four wings is lacking. Such insects would rapidly be selected out by natural selection (1).

"Control genes, like homeotic genes may be the target of mutations that would conceivably change phenotypes, but one must remember that, the more central one makes changes in a complex system, the more severe the peripheral consequences become. Homeotic changes induced in drosophila genes have only led to monstrosities," the evolutionist Schwabe admits (2).

In view of the ingenious way in which these main switch genes direct the subordinate genes of the morphogenesis (form creation) in a precise spatial and temporal pattern (3), it is difficult to attribute the origin of this symphony to random development over a long period of time. One false note, for instance one protein DNA interaction disrupted by a mutation, can, at any time (or with a temporal delay where there is buffering capacity), means the failure of the orchestra, (i.e., a maldevelopment or dysplasia) and thereby reduced health of both the individual and the species.

Every positive homeotic mutation, which is supposed to lead to a "higher" appearance, must be followed by a large number of small positive mutations in the target genes lower in the hierarchy of control. In purely mathematical terms, this brings about a multiplication of positive mutation probabilities. This drastically reduces the probability of realizing a new higher design process or phenotype by means of random mutations.

## References

1. Jane B. Reece und Neil A. Campbell. *Biology*. (Benjamin/Cummings, 1999): 460.
2. C. Schwabe. "Theoretical limitations of molecular phylogenetics and the evolution of relaxins." *Comp. Biochem. Physiol.*, 107B (1994): 167–177.
3. Walter J. Gehring. *Wie Gene die Entwicklung steuern*. (Birkhäuser Verlag, 2001).

# 12

# RUDIMENTARY ORGANS

In the last 150 years, in numerous creatures, organs have been discovered which were initially categorised as rudimentary, incomplete and useless. However, most often it was subsequently demonstrated that they were of great real benefit to the overall organism. In other cases it was down to atrophy. The billions of "organs under construction," with which nature should be awash, do not exist.

With evolutionary development through small mutational steps, one would expect that many creatures would be carrying with them, down the generations, organs that were half finished and in the developmental stage. Despite intensive search, no such thing has been found (1) (2) (3).

With closer study, most of the organs previously considered rudimentary have proven to be useful and beneficial (4) (5).

## Several examples:

- The human appendix assists the gut flora. In case of diarrhoea, part of the gut flora can survive in the appendix, which thereby helps to reconstitute the entire gut flora as quickly as possible. People who have had their appendix removed need longer to fully recover.
- The pelvic girdle remnants in whales have a relationship to the genital organs and serve as a point of attachment for the strong anal muscles.

- Embryonic tooth germs in baleen whales, which will never become real teeth, play an important role, as in all mammals, in the formation of the jawbones.
- The loss of eyesight in blind fish: a superfluous organ is broken down.

## References

1. Helmut Schneider. "Natura, Biologie für Gymnasien." Band 2, Lehrerband, Part B, 7. bis 10. Schuljahr, Ernst Klett Verlag. (2006): 268.
2. Prof. Ulrich Weber (Süßen). *Biologie Oberstufe, Gesamtband, Cornelsen Verlag.* (Berlin 2001): 259.
3. Horst Bayrhuber, Ulrich Kull, Linder Biologie, Lehrbuch für die Oberstufe, 21., neu bearbeitete Auflage, Schroedel Verlag GmbH, Hannover (1998): 404.
4. Junker, Ähnlichkeiten – Rudimente – Atavismen, 2002, Hänssler-Verlag.
5. Junker und Scherer, Evolution, ein kritisches Lehrbuch, Weyel (2006): 186–190.

# 13

# RECAPITULATION THEORY

Ernst Haeckel (1834-1919) asserted that the human being, while growing in the womb, repeats evolutionary development from fish to human. This thesis was already discredited in Haeckel's own lifetime. New photographs prove the complete soundlessness of this theory. In spite of this, Haeckel's depiction still appears in many schoolbooks!

Haeckel attempted to prove with drawings, that the vertebrate embryo, during its development, passes through stages of evolutionary development (1). However, these drawings were proved to be fakes. The fraud was already detected at the end of the 1860s (2) (3).

In 1997, the embryologist Richardson and his staff photographed various vertebrate embryos in different stages of development and arranged them in a similar way to that made by Haeckel. With the aid of these photographs, any layman can recognise how each type of vertebrate goes through its particular, individual developmental path taking the shortest route to becoming a viable individual (4).

It is quite incomprehensible how such a blatant falsification within a scientific work could have been propagated for over one hundred years, and that it can still be encountered in established teaching material to this day (5) (6) (7)!

## References

1. Ernst Haeckel, Natürliche Schöpfungsgeschichte, 1868.

2. Rolf Höneisen, Gefälschte Zeichnungen, factum (Januar 1999): 8–11.
3. Lee Strobel. Indizien für einen Schöpfer, Gerth Medien (2006): 42.
4. M.K. Richardson, J. Hanken, M.L. Gooneratne, C. Pieau, A. Raynaud, L. Selwood and G.M. Wright. "There is no highly conserved embryonic stage in the vertebrates." *Anatomy and Embryology* (1997): 196.
5. Helmut Schneider. *Natura, Biologie für Gymnasien*. Band 2, Lehrerband, Part B, 7. to 10. Schuljahr, Ernst Klett Verlag (2006): 277.
6. Horst Bayrhuber und Ulrich Kull, Linder Biologie, Lehrbuch für die Oberstufe, 21., neu bearbeitete Auflage. Schroedel Verlag GmbH, Hannover (1998): 402,406.
7. Prof. Ulrich Weber (Süßen). *Biologie Oberstufe, Gesamtband, Cornelsen Verlag*. (Berlin 2001): 257, 260.

# 14

# PEPPERED MOTH

In many schoolbooks, the peppered moth is cited as a prime example of observed evolution. There are two forms, one with lighter markings, and the other with darker ones. Due to air pollution during the Industrial Revolution, the white lichens on the bark of trees died off, and the trees were darkened. At that time, the dark moths proliferated faster than the light ones. This is supposed to have happened because the light moths could more easily be seen against the dark tree trunks by the birds that prey upon them. However, this process cannot even be termed micro evolution. It is simply a case of decrease/increase of existing populations.

After it had been assumed that, in the case of the peppered moth, one had found a real observable example of evolution, detailed field studies were carried out. And what did they prove?

Peppered moths hardly ever settle on tree trunks. In addition, the light form had already started to increase its numbers again, before the lichens had regenerated. Finally, it could even be demonstrated that these moths do not even tend to choose backgrounds matching their own colour.

## Evolutionary development

As far as alleged evolutionary development is concerned, only a shift of the allele frequency and not even the emergence of a new sub-species could be observed. This process can not even be tagged as micro evolution. Note also that the light forms have the dark brown pigment melanin, which is responsible for the colouring. Between the light and dark forms there is only a variation in the synthesis and distribution of melanin.

## Conclusion

Should there be any connection whatsoever between pollution and the frequency of dark and light moths? If so, then it is much more complicated than previously supposed, and so far, not understood (1). The fact that such an example should still be found in modern school textbooks (2) (3) clearly demonstrates how non-critically the theory of evolution in general is accepted.

## References

1. Junker und Scherer. *Evolution, ein kritisches Lehrbuch.* (Weyel, 2006): 71.
2. Helmut Schneider. *Natura, Biologie für Gymnasien*, Band 2, Lehrerband, Part B, 7. to 10. Schuljahr, Ernst Klett Verlag (2006): 270.
3. Horst Bayrhuber, Linder Biologie, Lehrbuch für die Oberstufe, 21. Auflage, Schroedel Verlag, Hannover, 388.

# 15

# DDT RESISTANT INSECTS

Flies and gnats becoming resistant, over a certain amount of time, to the insecticide DDT, were designated as proving evolution. However, subsequent studies have shown that insects resistant to DDT have always existed. All insects which are DDT resistant are descendants of these rare varieties. Simply, the non-resistant varieties have largely died out, while the resistant ones were able to continue to multiply.

The resistant flies and gnats in question go back to rare genotypes*, which did not fall victim to the initial mass deaths which occurred after the insecticide started to be used. The DDT resistant forms already existed before the use of the insecticide (1).

This example is not even micro evolution, as no new information entered the genes. No new traits came into existence; rather, there was only an extreme shift in the frequency of certain traits. In the process nothing new came into existence (2).

This alleged example of evolution is also found to this day in various different school textbooks (3) (4), although the interrelationships mentioned are acknowledged.

\* The genotype of an organism represents its exact genetic equipment and therefore the individual set of genes that it carries within itself in its cell nucleus.

# References

1. Junker und Scherer. *Evolution, ein kritisches Lehrbuch.* (Weyel, 2006): 73.
2. Lee Spetner. *Not by Chance!* (The Judaica Press, 1997): 143–144.
3. Helmut Schneider. *Natura, Biologie für Gymnasien*, Band 2, Lehrerband, Part B, 7. to 10. Schuljahr, Ernst Klett Verlag, (2006): 270.
4. Horst Bayrhuber, Linder Biologie, Lehrbuch für die Oberstufe, 21. Auflage, Schroedel Verlag, Hannover, 335, 364.

# 16

# RESISTANCE TO ANTIBIOTICS

The fact that bacteria can become resistant to antibiotics is often seen as an observable example of evolution. However, mutations which lead to a resistance to antibiotics will result, as a rule, in a loss of information in the genome. In the vast majority of cases, only a single base in the genome is altered, making it impossible for a certain bacterium to establish itself in the body of the host. In the process, there is no increase of information in the genome.

In a sufficiently large population, antibiotic resistant mutations on culture media containing antibiotics can be established particularly easily. Antibiotic resistant cells are, however, present even before the antibiotic takes effect. The antibiotic itself only fulfils a selection function. Lederberg's replica test (growth despite antibiotics) provides direct proof of this.

**An extract from** *Evolution, a Critical Textbook* **(1):**

"In order to understand the development of resistance on a molecular level, one should first of all observe antibiotics which inhibit protein synthesis by binding to ribosomal proteins. The resistance to antibiotic spectinomycin is associated with the structure of the S5 protein of the small ribosomal subunit. That is where the antibiotic binds. A mutation leads to an exchange of the amino acid serin with prolin at a particular place on the S5 protein. This exchange results in an alteration of the spatial structure of the protein, by which the binding site for spectinomycin is also affected. As a result, the antibiotic can no longer attack the S5 protein; the bacterium has become resistant.

A further possibility of the formation of resistance (e.g., to chloramphenicol) exists in detoxification by acetylation (binding of an acetic acid residue). This happens as a result of the enzyme chloramphenicol o-acetyltransferase (CAT) and is the result of gene duplication.

It is understandable that bacteria have mechanisms to degrade antibiotics, as fungi produce antibiotics naturally to use them for defence against bacteria.

## Penicillin synthesis (2)

The discovery of penicillin synthesis by brush-mould penicillium is a famous example. Penicillin inhibits the cell wall synthesis of bacteria and is split by resistant strains by penicillinase (beta-lactamase) and thereby rendered harmless. The gene for this enzyme is often localised on plasmids. One important class of antibiotic resistances is based on the new acquisition of genes by plasmid absorption (horizontal gene transfer).

There can be no doubt that the acquisition of antibiotic resistance is a micro evolutionary process with selection positive effect, if bacteria are exposed to antibiotics as a selection factor.

## References

1. Junker und Scherer. *Evolution, ein kritisches Lehrbuch.* (Weyel Verlag, 2006): 142.
2. Junker und Scherer, 143.

# GEOLOGY AND PALAEONTOLOGY

The model of an ancient Earth is of decisive importance to the theory of evolution. Only if the history of our planet is several billion years old, should it theoretically be possible for a simple monocellular organism to gradually evolve into a human being. The question nevertheless arises: is our Earth really billions of years old?

The so-called radiometric measuring methods, which are mainly used to determine the age of rocks and fossils, are by no means guaranteed. The available data can be interpreted in very different ways. More on that in the chapter Radiometry and Geophysics.

Various observations of geological formations leave huge room for doubt in traditional dating models. If one observes the erosion of the continents, the growth of river deltas and the changes to seacoasts and reefs, it is not conceivable that the actual processes have been going on for millions of years.

Examination of the layer boundaries between geological formations and knowledge of modern sedimentology both point to the history of the Earth being short. Catastrophic events such as the eruption of Mount St. Helens in the north western USA prove that the geological formations of our Earth could have been formed in a very short time span.

Finally, the fossil record contradicts Darwin's teachings on the origin of species. From an evolutionary theoretical perspective, many millions of intermediate forms must, by now, have lived on our Earth. Yet not a single undisputed evolutionary missing link has been discovered to date.

# 17

# STASIS IN THE FOSSIL REPORT

When Charles Darwin published his theory that all the creatures known to us are related to each other, he was rewarded with much enthusiasm from the majority of palaeontologists. Even then it could be recognised that the necessary transitional forms between the individual basic types were absent. Today, one can, on the basis of observation, talk of stasis as a main feature of the fossil record. Stasis signifies that no new forms or organs have come into being and the basic types have remained the same, in essence, for the whole of the Earth's history.

Due to the systematic lack of fundamental, directed changes in the fossils, the accepted higher development of beings must be regarded as a myth. In the development of most fossil types, there are two essential features which distinctly contradict slow development proceeding in small steps (gradualism): stasis and the sudden appearance of new species.

## Stasis

Most species demonstrate no targeted changes in the sequence of geological layers in which they occur. From the moment of their first appearance until their disappearance only limited and directionless changes are detectable.

## Sudden appearance of new species

Within the geological timetable, as a rule, new species appear suddenly and as fully developed. No fossils have ever been found which record the process of a gradual transformation from one species to another (1). Among the famous ammonites, several step-by-step changes are demonstrable. However, with these fossils, only the size and texture of the surface changed (micro evolution).

## Historical backgrounds (2)

"We palaeontologists have said that the history of life (the thesis of gradual transformation by adaptation) is underpinned by the fossils, while knowing all the time that basically, this was not the case," the famous palaeontologist Niles Eldredge pointed out. This is how, over the course of time, it developed into a palaeontologist's professional secret, that these evolutionary intermediary forms do not exist.

"It appears that every generation brings forth several young palaeontologists who are keen to document examples of evolutionary transition in the fossils. The alterations they looked for are, of course, supposed to be of a gradually progressing type. In most cases their efforts are not crowned with success. Their fossils appear to remain basically unchanged, instead of demonstrating the expected evolutionary forms," Eldredge states.

This extraordinarily high level of consistency in fossils appears to the palaeontologist who is determined to find evidence of evolutionary transition, as if no evolution had taken place. Indeed, because the basic concept of evolution is taken for granted, stasis is usually regarded as irrelevant to the results and the missing fossil transitional forms declared to be "gaps in the fossil record."

## Persisting species (3)

The term persisting species describes plant and animal species, which, throughout the entire geological period, have remained almost or completely unchanged. For instance:

- Viruses, bacteria and mould fungi since the Precambrian
- Sponges, gastropods and jellyfish since the Cambrian
- Mosses, starfish and worms since the Ordovician
- Scorpions and corals since the Silurian
- Sharks and lungfish since the Devonian
- Ferns and cockroaches since the Carboniferous
- Beetles and dragonflies since the Permian
- Pines and palms since the Triassic
- Crocodiles and tortoises since the Jurassic
- Ducks and pelicans since the Cretaceous
- Rats and hedgehogs since the Palaeocene
- Lemurs and rhinoceroses since the Eocene
- Beavers, squirrels and ants since the Oligocene
- Camels and wolves since the Miocene
- Horses and elephants since the Pliocene

On the basis of the evolutionary model, one expects the species to be in a permanent state of transition. Instead of this, as a rule, in all geological strata in which they appear, they are encountered unaltered. Cross connections between the species are entirely lacking.

## References

1. Stephen Jay Gould, quoted in Phillip E. Johnson's *Darwin im Kreuzverhör*, CLV: 66.
2. Niles Eldredge, quoted in Phillip E. Johnson's *Darwin im Kreuzverhör*, CLV: 76–77.
3. Willem J. Ouweneel. Evolution in der Zeitenwende, Christliche Schriftenverbreitung Hückeswagen: 146.

# 18

# RAPID FOSSILISATION (TAPHONOMY)

In order for a creature to become fossilised, it must be covered almost immediately with sediment and sealed off from the air. Otherwise it will rot /decompose. After the sealed creature has been surrounded by suitable minerals, due to the laws of chemistry an exchange takes place between the creature's molecules and its mineral-containing surroundings. The actual process can, under favourable conditions, actually begin within five days and be completed in weeks, months or a few years. How quickly a creature is mineralised depends on the surroundings in which it was embedded.

Fossils usually come about only as the result of major catastrophes. In the 1988 edition of Brockhaus, under the heading "fossilisation," it says the following: "The prerequisite (for the formation of fossils) is the rapid embedding of dead creatures in clay, sand and other sediments or in resin (later to become amber), so that they cannot decompose, be consumed or destroyed by any other external physical or chemical forces."

## Rapid fossilisation

According to a report by Derek Briggs and Amanda Kear in *Science* magazine, it has been observed in laboratory experiments that part mineralisation of shrimp occurred as soon as two weeks after death (1). There was already forty percent mineralisation of the muscles after only eight weeks. Even if this process does not always take place as quickly as that, it is, however, true that millions of years are not necessary.

## Dinosaur bones with elastic tissue and cellular structures

Interestingly, in past years some dinosaur bones have been found in which the process of mineralisation was not complete. Among other things, they contained elastic tissue with cellular structures (collagen and blood vessels). If one assumes that these bones are actually sixty million years old or older, then it is very hard to explain how this organic material could defy the decaying process (entropy) for such a long time (2) (3).

Furthermore, dinosaur bones have been found which contain fragments of protein. According to current knowledge, these should only be capable of being preserved for significantly less than one million years (4).

## Actualism and catastrophism

Actualism is one of the cornerstones of the theory of evolution. This doctrine says that, in the past, processes took place which are similar to those we still see today. Measures are therefore taken of the amount of material deposited each year at certain points on the seabed, and estimates made, according to this, of the time it would have taken to build up the whole stratum. To deposit a chalk stratum one metre thick would, under the present environmental conditions, take approximately 40,000 years. It also, however, has to be borne in mind that fossils of soft body parts and plants could come about only if the creature was buried quickly and completely enough so that neither air and water nor bacteria and scavengers could damage it.

Most rock strata found today contain larger or smaller fossils. All these strata must have been built up very quickly.

In Sweden, half of the Ordovician (allegedly some thirty million or more years old) can be viewed in a single quarry. This is called a condensation camp because one assumes that the stratification occurred very slowly. However, even in these deposits, any number of trilobites can be found (5). These stratifications must have taken place in phases, which could have occurred within days, years or decades. Otherwise the trilobites would have disintegrated before they could fossilise.

## References

1. Derek E.G. Briggs and Amanda J. Kear. "Fossilization of Soft Tissue in the Laboratory." *Science* 259 (5 March 1993): 1439–1442.
2. Mary Higby Schweitzer, et al. "Analyses of Soft Tissue from Tyrannosaurus rex suggest the Presence of Protein." *Science* 316 (13 April 2007): 277–280.
3. H. Binder. Elastisches Gewebe aus fossilen Dinosaurier-Knochen, Studium Integrale, (Oktober 2005): 72–73. http://www.wort-und-wissen.de/index2.php?artikel=sij/sij122/sij122-5.html.
4. H. Binder. "Proteine aus einem fossilen Oberschenkelknochen von Tyrannosaurus Rex." *Studium Integrale* (October 2007): 78–81 (A.d.L.: The spelling "Tyrannosaurus Rex" appears mainly in German publications, while the variant Tyrannosaurus rex is to be found published in English).
5. R. Fortey. Trilobiten! (München, 2002): 203.

# 19

# MISSING LINKS

The required transitions from fish to amphibians, from amphibians to reptiles and from reptiles to birds, have, even after 150 years of research among the fossils, not been found. Comparisons between the "most amphibian like fish" (coelacanth/periophthalmus) and the "most fish-like reptiles" (ichthyostega) demonstrate, moreover, that with complex key attributes such as the construction of the tetrapod extremities (legs of the four-footed land creature) or the construction of the skull, evolutionary intermediate forms are hardly conceivable. For the transition between reptiles and birds, the idea that archaeopteryx was a transitional form is stubbornly held, even though it has been proved that it was one hundred percent a bird, feathered, warm blooded and equipped with the special bird lung.

Between the different orders, families, and classes of creature known to us and passed down in fossils, there is not a single undisputed transitional form. Between all these classes and their many orders, one would, in accordance with the theory of evolution, expect countless transitional forms, combining the many key traits of two species. Several examples have, in the past, been suggested as transitional forms; however, after extensive examination, all were rejected (1) (2) (3).

## The coelacanth (crossopterygian)

The coelacanth is supposed to be one of these transitional forms between fish and amphibians. This fish has fins with enhanced muscle attachments so that one assumed that it would walk on the seafloor with its fins. These creatures were observed day and night and, as a result, the conclusion was reached that it uses its enhanced muscle attachments to stand upright in the water and to swim upright with its head upwards and its chest forwards. Practically no textbook mentions anything about this.

If one observes the coelacanth (for instance, latimeria, a living fossil) it becomes clear that it is undoubtedly a fish. Added to that is the fact that, at one metre long, it is a relatively large fish. The assumption that this large fish should be a transitional form between fish and amphibian is not at all credible. Furthermore, it lives at great ocean depths and there is no trace of even the beginnings of lungs.

## Archaeopteryx

Since the discovery of archaeopteryx in the 1860s, the phylogenetic origins of the bird have been discussed controversially (4). The central question often concerned its ability to fly, especially with respect to its suspected descent from a dinosaur that walked on two legs (theropods, for example compsognathus; according to later opinion, thecodonts) (5).

Based on the earlier anatomical-morphological studies carried out as early as in the nineteenth century by the biologist Thomas Huxley, this idea has been taken up repeatedly by taxonomists and palaeontologists up until the most recent times. A good flying ability, resulting from therapod descension, is doubted (6).

It's true that even the palaeornithologist Alan Feduccia does not fundamentally rule out the descent of birds from the dinosaurs of the tree dwelling/living kind, capable of flying or at least gliding (7). However, contradictory discoveries, for instance concerning the identity of morphological structures (bird hand bones), complicate the interpretation of phylogenetic connections. With the assistance of familiar fossil evidence, far and wide no dinosaur forerunner is in sight, which could be deemed to be the progenitor of all birds.

The fact that this one disputed form is, time and again, cited as an example of transitional forms illustrates how poorly stocked the number of known transitional forms is. Nevertheless, one must be aware that the development of wings capable of supporting flight poses a very special problem for the idea of evolution proceeding over many generations: feathered wings, a bird's heart and a bird's lungs offer the creature a survival advantage only if they are complete and fully functional.

## The snake

The snake's family tree can only be recognised in fossils in a very fragmentary way, if at all. Among experts, the evolution of today's snake is a phenomenon which can only be explained by much speculation (8).

## The mudskipper (periophthalmus)

At the first glance, one could take the mudskipper for a transitional form between fish and amphibian, yet well-known evolutionary researchers seldom believe that. In spite of its amphibian way of life, fins and breathing gills show that it is a fish. With mudskippers, the gill cavity is only connected to the outside world by a narrow gill slit, which prevents the delicate breathing organ from drying out. By snatching air it can, within limits, refresh the oxygen content of the seawater it holds in its enlarged jaw cavity (9).

# References

1. Helmut Schneider. *Natura, Biologie für Gymnasien*, Band 2, Lehrerband, Part B, 7. to 10. Schuljahr, Ernst Klett Verlag (2006): 257.
2. Horst Bayrhuber & Ulrich Kull. *Linder Biologie, Textbbok for senior classes*, 21. (Schroedel Verlag GmbH, Hannover, 1998): 418, 430, 432.
3. Ulrich Weber. Biologie Oberstufe, Gesamtband, Cornelsen Verlag, (2001): 294–295.
4. Helmut Schneider. *Natura, Biologie für Gymnasien*, Band 2, Lehrerband, Part B, 7. to 10. Schuljahr, Ernst Klett Verlag (2006): 261.
5. G. Heilmann. *The Origin of Birds*. (London: Witherby, 1926).
6. R.T. Bakker. "Dinosaur renaissance." *Scientific American* 232 (1975): 58–78.
7. Alan Feduccia. "The problem of birds origin and avian evolution." *Journal Ornithology* 142, Sonderheft 1139–1147, (Studium Integrale, Mai 2002: 37–40).
8. Colbert et al. *Evolution of the vertebrates: A history of the backboned animals through time* 5. edition (New York: Wiley-Liss, 2001).
9. P.K.L. Ng und N. Sivasothi. *A Guide to the Mangroves of Singapore* 1. (Singapore Science Centre, 1999):138–139.

## 20

## CAMBRIAN EXPLOSION

In the Earth's layers that predate the so-called Cambrian (which allegedly took place 488 to 542 million years ago), one finds only microfossils. In the Cambrian itself highly differentiated creatures suddenly appear. The assumption that single-cell and multi-cell organisms or plants and animals have common ancestors is not supported by the fossil record but rather massively challenged. This problem is generally acknowledged. Because the higher creatures appear explosively and without forerunners, specialists talk about the "Cambrian explosion."

The lowest Earth stratum to unequivocally contain fossils is referred to as the Cambrian. Cambrian explosion means the sudden appearance of many new blueprints, allegedly some 530 million years ago (1).

Eighty-seven percent of all phyla (plants and animals) appearing in the upper layers have also already occurred in the Cambrian. Only the vertebrates, the bryozoans and the insects first appear in the Earth's upper layers (the Ordovician and/or Devonian). In the Earth's strata older than the Cambrian, hardly a single undisputed higher fossil appears. There is, therefore, not a single piece of fossil evidence that the creatures that appear in the Cambrian explosion have common predecessors.

According to the theory of evolution, within the extremely short time of allegedly five to ten million years, at least nineteen to thirty-five new phyla (from a total of forty) appeared on the Earth for the first time (2) (3). Many new sub-phyla (thirty-two to forty-eight of a total of fifty-six) and classes of animals likewise appeared in these strata for the first time. All representatives of these phyla have important morphological traits. The morphological antecedents expected, according to the theory of evolution, in the earlier Vendian or in the Precambrian fauna are absent in almost every case (4).

More recent discoveries and analyses show that these morphological gaps are not accounted for simply by citing an incomplete fossil history (5). Because it is assumed that the fossil history is more or less reliable, scientists debate whether this observation concurs with the strictly monophyletic (a single comprehensive family tree) view of evolution (6).

## Fast or slow fuses

Those who believe fossils offer a reliable picture of the appearance of the so-called metazoans lean towards the view that these animals came into existence relatively quickly. Therefore, the Cambrian explosion had a so-called "fast-fuse" (7). Some (8), but not all (9), who think that the molecular phylogenies provide more reliable branching times of the Precambrian antecedents, believe that the Cambrian animals developed over a much longer time period and that the Cambrian explosion therefore had a "slow fuse."

Ernst Mayr, the main advocate of the modern synthetic theory of evolution, who died in 2005, said when asked about the Cambrian explosion (10):

"Almost all ... phyla appear at the end of the Precambrian and the beginning of the Cambrian, that is, some 565 to 530 million years ago, in fully developed form. No fossils have been found which stand between them, and even today there are no such intermediary forms. The phyla appear, therefore, to be separated by unbridgeable gaps."

# References

1. Junker und Scherer. *Evolution, ein kritisches Lehrbuch.* (Weyel, 2006): 227.
2. Ernst Meyer et al. "DNA and the origin of life: information, specification and explanation," in J.A. Campbell und S.C. Meyer, *Darwinism, Design and Public Education.* (Michigan State University Press, 2003): 223–285, http://www.discovery.org/scripts/viewDB/index.php?command=view&id=2177.
3. S.A. Bowring, J.P. Grotzinger, C.E. Isachsen, A.H. Knoll, S.M. Pelechaty und P. Kolosov. "Calibrating rates of early Cambrian evolution." *Science* 261 (3 September 1993): 1293–1298.
4. G.L.G. Miklos. "Emergence of organizational complexities during metazoan evolution: perspectives from molecular biology, palaeontology and neo-Darwinism." Mem. Ass. Australas. Palaeontols 15 (1993): 7–41.
5. M. Foote. "Sampling, taxonomic description and our evolving knowledge of morphological diversity." *Paleobiology* 23 (1997): 181–206.
6. Simon Conway Morris. "The question of metazoan monophyly and the fossil record." *Progress in Molecular and Subcellular Biology* 21 (1998): 1–9.
7. Simon Conway Morris. "Cambrian 'explosion' of metazoans and molecular biology: would Darwin be satisfied?" *International Journal of Developmental Biology* 47 (2003): 505–515.
8. Gregory A. Wray, Jeffrey S. Levinton und Leo H. Shapiro. "Molecular Evidence for Deep Precambrian Divergences Among Metazoan Phyla." *Science* 274 (25 Oktober 1996): 568–573.
9. Francisco José Ayala, Andrey Rzhetsky und Francisco J. Ayala. "Origin of the metazoan phyla: molecular clocks confirm paleontological estimates." *Proc Natl Acad Sci USA* 95 (20 Januar 1998): 606–611.
10. Ernst Mayr. *Das ist Evolution.* 3. A., (München, 2003): 74.

# 21

# EROSION OF THE CONTINENTS

The renowned geologist Ariel A. Roth researched how much rubble, mud, and debris today's rivers wash into the oceans, year after year. He calculated that after ten million years, the continents would be eroded to sea level if they were not simultaneously being raised by tectonic processes. Even if, in the past, significantly less material had been washed down, it is clear that at least in the upper rock strata, it should not have been possible to find fossils that were considerably older than ten million years. They must have long since been washed away.

Today, the Earth's continents stand, on average, 623 metres above sea level. They are being constantly worn down, mostly by the effects of rain and washed into the oceans by rivers and streams. At the current amount of these transported materials, it would take some ten million years to wear all the continents down to sea level. In "only" 185 million years, the material carried down would correspond to the volume of the present day oceans (1).

## Consequences for the geological timetable

As our continents are in the process of such a powerful transition, it is inconceivable that the fossils that we find on the Earth's surface should actually be 300–500 million years old. They could not have been found in the abundance we find them today.

The conventional geological timetable, as is taught in most state schools, has to be examined very critically.

## Flood disasters of global proportion

This is further complicated by the fact that, in the above calculation, no account has been taken of the fact that, in the past, one or more flood disasters of global proportion have taken place. We can see this from numerous geological finds. A lot of additional material would have to be washed into the sea by a global flood.

## Reference

1. Ariel A. Roth. "Some Questions About Geochronology." *Origins* 13 Nr. 2 (1986): 65.

# 22

# RIVER DELTAS, SEA COASTS AND REEFS

The material washed into lakes and seas by rivers and streams allows conclusions to be drawn as to how long these processes have already lasted. It is amazing that there is not a single river delta anywhere on Earth that can definitely be significantly more than 10,000 years old. Even if one studies the current changes in lakes and sea coasts, the Earth's surface, as it is today, can never be millions or billions of years old.

The Amazon transports 500 million tons of material per year into the Atlantic. As a result, the shelf in the estuary area of the Amazon is raised every year by about 50 m above the surrounding shelf. Under today's conditions, it would have taken about 14,000 years to deposit this volume. If it is considered that the Amazon washed much more material into the sea during the formation of the Andes than it does today, then this figure must again be drastically reduced. In a good 3,000 years, today's shelf off the estuary of the Amazon will be filled up to the sea's surface.

The Mississippi transports around 300 million tons of material yearly into the Gulf of Mexico. On the basis of this volume, the gulf would have to be completely filled after eight million years. In truth, however, there is only a relatively small river delta of 50 km in length at the end of the river. Some advocates of a billion year old world say that the material washed down sinks continuously into the sea. Yet there is no trace of it in the drill cores that have been made in the seabed. Such drillings have been made all over the gulf, not to calculate the age of the Earth, of course, but in an attempt to find oil.

The Niagara Falls cliffs, due to the large volume of water, eroded 1.5 m each year. The waterfall is therefore shifting gradually in the direction of Lake Erie. From the distance from Lake Ontario (11.5 km) one can conclude that the Niagara Falls are, at the very most, 10,000 years old (1).

On England's Atlantic coast material is constantly being eroded due to the action of the waves. As a result, the coastline is moving inland at an average rate of one metre every six years. That means that after only a few million years, England would have completely disappeared. Interestingly, over the whole island group, countless fossils are to be found, which, by conventional reckoning, are supposed to be several hundred million years old (2). This timetable must be viewed critically.

In North Carolina (USA), in certain places the sea is eating away up to 4.2 m of land per year. On the other hand, the ancient town of Ephesus in present day Turkey was still, up until less than 2,000 years ago, a seaport, while today it lies several kilometres inland. These shifts underline the pace of geological events (3).

On the basis of the amount of gravel and sand being introduced, the Vierwaldstättersee (Lake Lucerne) in Switzerland will be completely filled in 4,000 years at the most. The Bodensee (Lake Constance) will not exist for more than 10,000 years from now.

Also in respect of the growth of limestone reefs, the consequences of a "million-year-old Earth" history are completely missing. The growth of the Zechstein reefs can be explained. Today, on the banks of the Bahamas, limestone particles are washed up, forming microbial mats; lamellae of approximately one millimetre per day are formed (4). There are approximately fifty lamellae in 17 mm of carbonate; using a simplified calculation, that only adds up to around 500 years development time for the highest reef (some 60 m) in Thuringia (5).

Until recently, the Carbonates of the Upper Jura (also called the White Jura or Malm), which have grown up to 200 m in height, were predominantly interpreted to be sponge reefs. Nowadays no comparable reef structures are known outside of the seas. Disbelief in this suspicious uniqueness and new studies carried out in the course of the last two decades have led to new approaches in reef limestone research. One can assume that, in the future, sedimentation processes will be increasingly used for model conceptions of the formation of reef limestone. Concerning the time question, carbonate sands as sediments, especially in energy-rich moving shallow water, form substantially faster than grown reef structures (6).

## References

1. Larry Pierce. "Niagara Falls and the Bible." *Creation* 22 (2000): 8–13. http://www.creationontheweb.com/content/view/276/.
2. A. Phillips and Tall Order. "Cape Hatteras Lighthouse makes tracks." *National Geographic* 197 (2000): 98–105.
3. Tas Walker. "Vanishing Coastlines." *Creation Ministries Magazine* 29, No. 2 (March to May 2007): 19–21.
4. C.D. Gebelein. "Distribution, Morphology and Accretion Rate of recent sub tidal Algal Stromatolites." *Bermuda, Journal of Sedimentation and Petrol* 39: 49–69.
5. K. Kerkmann. Riffe und Algenbänke im Zechstein von Thüringen, Freiberger Forschungshefte (1969): 252.
6. M. Stephan. Neue Interpretation der Massenkalke des süddeutschen Oberjura, Studium Integrale (October 2001): 91–94. http://www.wort-und-wissen.de/index2.php?artikel=disk/d08/4/d08-4.html.

# 23

# ERUPTION OF MOUNT ST. HELENS

In the course of the gigantic eruption of the Mount St. Helens volcano in the year 1980, within hours and days geological formations were created which correspond very closely to others which, up to now, were thought to have been formed in a process taking thousands to millions of years. The observations of Mount St. Helens illustrate the fact that the geological formations of our Earth could have been formed in a series of short catastrophic events.

Before the eruption in 1980, Mount St. Helens, located in north western USA, was approximately 400 m higher than it is today. As a result of the heat generated by the eruption, the snow in the area of the summit of the nearly 3,000 m high mountain, mingled with sediment and rock debris. The streams of mud and debris flowed down into the valley at speeds of up to 150 km/h and, within a short amount of time, eroded canyons up to 200 m deep in the solid rock.

Regarding other canyons in America, most geologists assume that they were slowly carved out by the water of rivers during very long periods of time (slow erosion). The eruption of Mount St. Helens proves, however, that such geological formations can develop in a very short amount of time.

As a result of the explosion, around a million tree trunks were hurled into the nearby Spirit Lake. New canyons and new river systems and lakes were created and the water level of Spirit Lake was raised by approximately 75 m (1).

After the eruption, the lake was covered with an immense mat of Douglas fir, noble fir, hemlock fir, silver fir, western red cedar and Alaska yellow cedar. Careful observations have shown that the trunks tend to float in an upright position with the roots downwards. In the course of time the trees sank and were deposited on the lakebed. Some of the trees embedded themselves upright on the lakebed.

If we were to find these trunks in fossilised form within rock strata, they would appear to us to be a naturally-growing forest. It would seem that a forest of noble firs was followed by a forest of hemlock firs and finally by a forest of Douglas firs. The buried forests in the Ruhr Coal Basin can be cited as an example from the historical past. At that time, many tons of cortex trees up to 12 m high during the Carboniferous period were completely buried in the mud (2).

## The formation of peat and coal

The waves in Spirit Lake caused friction between the tree trunks. This caused the water-soaked bark pieces to break off from the trunks, gradually covering the lakebed. Thus, within a few years, a peat layer, consisting of up to twenty-five percent tree bark and being several centimetres thick, was created. Studies showed that this peat has a close structural relationship to brown coal. Perhaps in Spirit Lake we are witnessing the first stage of the formation of coal.

## References

1. Wort und Wissen. Diaserie Ausbruch des Mt. St. Helens, zu finden unter http://www.wort-und-wissen.de/index2.php?artikel=medienstelle/diaserie.html.
2. H. Klusemann und R. Teichmüller. Begrabene Wälder im Ruhrkohlenbecken, Natur und Volk 84 (1954): 373–382.

# 24

# MODERN SEDIMENTOLOGY

Modern sedimentology confirms that the characteristic traits of the sediment strata, which are visible and accessible to researchers, point to short and intensive depositions. It would be difficult to validate the time periods of up to hundreds of millions of years on the basis of the observed sedimentary structures (cross bedding/ graded bedding). A rethinking is also taking place with regard to the interpretation of ultra-fine strata. In many cases, people are talking in terms of days rather than years.

On May 24, 2002, at the International Solomon University in Kiev (Ukraine), the question; "Is macro evolution and progressive evolution fact?" was debated for the second time. Ten different speakers gave their views on the subject. Especially worth mentioning, is the geologist A.V. Lalomov, director of the geological research laboratory ARCTUR in Moscow, who advocated a "short time geology without compromise."

In his opinion, modern sedimentology confirms that the true traits of the sedimentary strata that are visible and accessible to research (as opposed to the gaps between the strata which provide nothing observable or researchable) demonstrate short and intensive depositions (1).

It is interesting that, for several years, Russian scientists (primarily those in search of natural resources) have increasingly started to doubt the model of a billion-year-old Earth. It is interesting because the former Soviet Union was a bastion of atheistic evolutionary thinking.

## Cross bedding

The structures of deposited sediment allow clues to the speed with which they were deposited. Cross bedding occurs by way of fast-flowing water, regardless of how big the areas are. The more water involved in this process, the deeper the layers created. Cross bedding can be observed from an extent of a few centimetres, up to a depth of 20 m. A large proportion of sediment throughout the world is cross-bedded.

## Graded bedding

At the lowest level, graded bedding strata contain coarse material, which becomes finer towards the top. Graded bedding layers must come into existence within hours, days and weeks. They are created during the abating of a flood in the course of which the speed of the water slowly diminishes. Coarse material is moved and deposited at higher water speeds, fine material at lower speeds. Another large proportion of sedimentary layers throughout the world is graded bedded.

## Depositing of finest strata

The scarcely millimetre thick paper shale in the Saar-Pfalz Permian strata in Germany can, according to recent findings, be interpreted as light silts and dark clays with organic matter, which are produced on an alternating daily basis. Daily storms produced turbidity sediments and graded silts (2).

## Missing deposition interruptions

The Schmiedefeld Formation (Ordovician from Thuringia) is, for example, conventionally quoted as having taken twenty million years to come into existence (3). However, as regards to this formation, no long deposition interruptions are discernible, but rather indications of continuous, even relatively rapid sedimentation. That leads, in long-term understanding to a scarcely resolvable contradiction between rapid sedimentation (high sedimentation rate), a low overall depth and a long development time. The findings indicate a development time of only centuries rather than millions of years (4).

## References

1. Alexander Lalomov, et al. "Soviet scientists and academics debate Creation-evolution issue." *Technical Journal* 17/1 (2003): 67–69.
2. Andreas Schäfer, Klastische Sedimente, München (2005): 171f.
3. M. Menning & Deutsche Stratigraphische Kommission, Stratigraphische Tabelle von Deutschland 2002, Potsdam, 2002.
4. J. Ellenberg. Die Bildung oolithischer Eisenerze im thüringischen Ordovizium, Geowiss. Mitt. v. Thüringen, Beiheft 9 (2000): 57–82.

# 25

# UNDAMAGED LAYER BOUNDARIES

The layer boundaries for geological formations (the transition from one sedimentary layer to the next), which are often attributed an age of thousands or more years, generally show no or little surface erosion, bioturbation or formation of soil. The proposition that the surface of an Earth's stratum escaped the effects of the weather for millennia, before being covered by another layer, is inconceivable. Therefore the majority of sedimentary strata must have come into existence within days, years and decades.

The vast majority of sediment layers are either obliquely layered, graded and/or contain fossils. In the formation of the sediment layers themselves, an interpretation of millions of years can never be given. It must be clarified what period of time could have elapsed between the coming into existence of one layer and the next.

The following characteristics show a rapid process of layering, they are characteristic of most geological layer boundaries:

a) Insufficiently eroded surfaces (1): if a surface is exposed to the elements for a long period of time, it becomes eroded. Water and wind, by the action of erosion, form uneven and indented surfaces. The longer the weather has an effect, the more distinct the surface irregularities and indentations become. After only a few decades, as a rule significant alterations to the surface can be observed. How have the layer boundaries in geological formations, which allegedly boast an age of tens of thousands of years, remained largely undamaged?

b) Little or no bioturbation (2): On a seabed or lake bed, after a certain time plants and animals start to establish themselves and leave traces behind: root formation by plants, signs of digging by burrowing molluscs and other digging creatures, wormholes, etc. If such a surface is covered with sediment, the traces remain. If such traces are missing or sparse, then one has to assume that these layers were deposited in rapid succession.

c) Soil formation: complicated chemical processes lead, in the course of a few hundred years, to the formation of soil. Soil formation is clearly visible on the immediate surface, while the characteristic traits of the formation of soil are scarcely seen in deeper geological layers. As iron oxide is found in all fertile soils, at least one black or brown colouration must be identifiable. Most of the deeper lying layers had to have been laid down so quickly that there was too little time for the formation of humus to take place.

d) Animals' footprints: One mainly finds footprints in geological layers of volcanic ash. Volcanic ash hardens very quickly. If it becomes moist and is dried out by the sun, the surface solidifies with the footprints. When covered with new material, footprints are preserved in clay, sand and other soft surfaces. Even if very few layer boundaries contain footprints, one can still assume that at least the layers containing the footprints came into existence very rapidly.

# References

1. Joachim Scheven. Karbonstudien, Hänssler-Verlag, (1986): 71.
2. Eugen Seibold and Wolfgang H. Berger. *The sea floor*. (Springer Berlin, 1996).

# 26

# POLYSTRATE FOSSILS

Time and again one finds fossil tree trunks, plants and animals, which extend through several geological layers (Polystrate fossils). The problem is that these layers are often attributed a radiometric age difference of several thousand or even tens of thousands of years. Thus, on the Hauenstein (in Switzerland) an ichthyosaur fossil, which extended through three layers, was found. Since fish start to decompose within a few days, layers such as the one on the Hauenstein must have formed very quickly. A tree trunk must be encased within a few years, or decades, for it to become petrified; otherwise it disintegrates.

The article "Hauenstein ichthyosaur puzzle solved?" in the *NZZ (New Zurich Newspaper)* on March 12, 2004, is an example of how the model of a billion-year-old Earth is taken for granted in the public media. The attempted explanation, that the ichthyosaur fossil of Hauenstein was retroactively impelled through several layers by an internal explosion, is somewhat strange (1). How could the carcass have remained completely undamaged?

"Unusual means" in interdisciplinary research and collaboration provided the answer to the embedding and fossilisation of the "Hauensteiner Dickschädel." Apart from the classic methods of geology and palaeontology, forensic and veterinary medicine, gynaecology, marine biology, computer tomography, submarine technology as well as other methods were employed (2).

The original paper by Hannes Hänggi and Achim G. Reisdorf has certainly earned recognition (3). Assuming that the ichthyosaur really is 190 million years old and that the layers through which it extends must be ascribed an age difference of more than just a few days, they have worked out a valid explanation.

Which of the following explanations is yet, in the final analysis, more plausible?

a) The carcass was retroactively impelled through several layers and remained undamaged, or
b) The limestone layers in which the fossil was found were laid down in a very short amount of time.

Because Polystrate fossils, such as this fish, are by no means a rarity, all the common timetables should be critically assessed.

Especially in coal-bearing layers, one finds carbonised or petrified Polystrate tree trunks (4), which illustrates that the geological layers in which they are found probably came into existence much more rapidly than estimated by conventional geology (5).

## References

1. Rolf Höneisen. Den Kopf im Fels, factum März 2004, http://www.factum-magazin.ch/wFactum_de/natur/Palaeontologie/Hauenstein_Ichthy.php.
2. A. Niederer. Ichthyosaurier-Rätsel vom Hauenstein gelöst? Der Weg des Schädels durch drei Gesteinsschichten, NZZ (Neue Zürcher Zeitung), Nr. 60 (12 März 2004): 19.
3. Hannes Hänggi und Achim G. Reisdorf. Der Ichthyosaurier vom Hauensteiner Nebelmeer - Wie eine Kopflandung die Wissenschaft Kopf stehen lässt, http://www.ngso.ch/06_Publikationen/PDF/120312_Saurier_7_22.pdf

4. Michael J. Oard und Hank Giesecke, Polystrate Fossils Require Rapid Deposition, CRSQ 43 (4 März 2007): 232–240.
5. Joachim Scheven. Karbonstudien, Hänssler-Verlag (1986): 31–41.

# 27

# LIVING FOSSILS

Most basic types of the animal and plant world are found in fossils. Those species, which are found in deeper rock and are completely absent in the layers that follow, although some reappear in upper layers and are still alive today, are called living fossils. The existence of living fossils raises doubts as to the reliability of the current interpretations of the fossil record. Should different geological ages really be ascribed to the individual geological layers in which living fossils appear? The numerous finds of living fossils call this interpretation into question.

In past years, as more living fossils have been discovered, experts searched for possible explanations. Advocates of the theory of evolution concluded that these numerous living fossils have survived for a certain period of time in "geologically non-traditional habitats" (1).

"Geologically non-traditional habitats" is open to a wide range of interpretations. In addition, all the missing links which are supposed to unite the parallel running lineages of the fossil record to a single family tree could - indeed must - have developed even further over millions of years in these habitats.

## Examples of living fossils in the plant kingdom:

- tree ferns (Cyatheales)
- the Ginkgo ("Temple Tree," Ginkgo biloba)

- the Cathaya (argyrophylla)
- the Welwitschia (Welwitschia mirabilis, a naked-seed desert plant)
- the Wollemia (Wollemia nobilis, of the family Araucariaceae )
- the Metasequoia (Metasequoia glyptostroboides)

## Examples of living fossils in the animal kingdom:

- the Alligatorfish (Cociella crocodila)
- the Tuatara (Sphenodon punctatus)
- the Manjuari (Atractosteus tristoechus, a garpike)
- the Purple Frog (Nasikabatrachus sahyadrensis)
- Neopilina galatheae (a mollusc)
- the Nautilus (Nautilidae, the earliest form of cephalopod)
- the Horseshoe Crab (Limulidae)
- the Coelacanth (Latimeria)
- the Lamprey (Petromyzontidae)
- Monotremes (Monotremata: the spiny anteater, Protheria, and the platypus, Ornithorhynchus anatinus
- Triops, a brachiopod (2)
- the Devil's Hole Pupfish (Cyprinodon diabolis)
- the Solenodon (Solenodontidae)

## References:

1. W.J. Ouweneel. Evolution in der Zeitenwende, Christliche Schriftenverbreitung Hückeswagen, 148.
2. Joachim Scheven. Null Evolution: Der Kiemenfuß, Leben Nr. (6 Januar 1995): 13.

# 28

# MILLION-YEAR-OLD ARTEFACTS

Time and again, some objects which are found in the Earth's layers and were with great probability made by human beings, so-called artefacts, are according to conventional estimations supposed to be far more than 100 million years old. Much speculation has been generated by these finds. Is humanity much older than previously thought? Do certain objects originate from aliens? Are we dealing here with time travellers? Only one thing is hardly ever questioned: the reliability of current geological timetables.

In June of 1934, a piece of wood was found sticking out of a piece of limestone. When an attempt was made to free the piece of wood, it was established that it was a wooden hammer handle. At the time of this discovery, the object was completely encased in limestone. The conclusion that must be drawn from this is that the hammer must have been made before the rock was formed. The age of the rock is estimated by geologists to be sixty-five to 140 million years old (1).

In the year 1989, an analysis of the metallic hammer head was undertaken. It is astonishing that there were no traces of carbon or any other additives in the hammer head. On the other hand, chlorine and sulphur were discovered. Nowadays, no process for producing iron is known which introduces these additives. That confirms the assumption that this hammer was manufactured before the modern Iron Age.

In their book "The Hidden History of the Human Race," M.A. Cremo and R.L. Thompson described, among other things, fifty-eight different man-made objects and human bones found in geological layers. The age of some of these, according to conventional evaluation, was estimated to be far above 100 million years.

## Further examples (Page indications refer to Reference 2)

- An iron nail embedded in Scottish sandstone, which is supposed to be between 360 and 408 million years old. (105)
- A beautifully decorated metal vase in Dorchester, Massachusetts embedded in a formation, ascribed an age of more than 600 million years. (106)
- A metal pipe in a piece of limestone, which was found in the quarry of Saint-Jean-de-Livet in France, has a specified age of sixty-five million years old. (117)
- A small golden chain, encased in coal that has been dated at 260 to 320 million years old. It was found in a mine in northern Illinois (USA). (113)
- A metal ball with grooves around its widest diameter, found in a layer of Pyrophyllite in South Africa. This layer has been dated at 2.8 billion years old. (121)
- A large number of different stone tools found in Boncelles, Belgium, in a layer with an age of twenty-five to thirty-eight million years old. (68–70)
- A golden thread embedded in stone in a quarry in Rutherford, England, with the given age of 320 to 360 million years old. (106)

Such finds must not be overrated. They do, however, cast justifiable doubt on current dating methods.

## References:

1. *Washington Post*, 2 Oct. 2005.
2. M.A. Cremo and R.L. Thompson. *The Hidden History of the Human Race*. (Govardhan Hill Publishing, Badger, USA, 1994).

# 29

# MILLION YEAR OLD MICROBES

It is not unusual for viable microorganisms to be found in salt and coal deposits, which are allegedly up to 500 million years old. A great deal of documentation exists on the isolation and reactivation of such microbes. These microbes could well be a few thousand years old, but in no way hundreds of millions of years old. Over such a length of time, the DNA and other cell building blocks would have long since decomposed. It is inconceivable that microbes in a frozen sleeping state (Cryptobiosis), and without nourishment, could be able to regenerate and repair themselves for such a long period.

Microorganisms occur practically everywhere in nature. Due to their extreme flexible physiology, they colonise an unprecedented variety of habitats. They are found in volcanic vents, in hot springs both on the Earth's surface and on the deep ocean floor, in the ice, in the Dead Sea and as symbionts (i.e., in the digestive tracts of higher organisms).

In recent years, however, it has not been unusual to find microbes in old salt and coal deposits. Many of these sites are attributed to the Permian (250 to 300 million years ago) or the Upper Precambrian period (up to 500 million years old).

Under the strictest security conditions, due to the danger of contamination by present-day microbes, various teams have been able, in various laboratories, to reactivate so-called ancient microbes out of their sleeping state and cultivate them (1).

It is even disputed among advocates of a billions-of-years-old Earth whether or not these microbes are really hundreds of millions of years old. The following points of criticism are usually cited:

a) The age of the isolated microorganisms can only be indirectly determined via dating of the matrix in which they are enclosed. In this case, diffusion cannot be ruled out as a source of error.
b) The possibility of contaminating recent microorganisms (i.e., those living today) when taking or processing samples cannot, even under the strictest laboratory conditions, be ruled out.

Normally, however, all known sources and possibilities of contamination are investigated according to the publications. Critics should cite concrete, non-checked possibilities of contamination. Across-the-board accusations of contamination lose credibility in the face of the wealth of data presented (2).

## References

1. Russel H. Vreeland, William D. Rosenzweig und Dennis W. Powers. "Isolation of a 250 million-year-old halotolerant bacterium from a primary salt crystal." *Nature* 407 (19 October 2000): 897–899.
2. Harald Binder. "Dornröschenschlaf bei Mikroorganismen?" *Studium Integrale* (October 2001): 51-55. http://www.wort-und-wissen.de/index2.php?artikel=sij/sij82/sij82-1.html.

# 30

# NUSPLINGEN PLATY LIMESTONE

Until a few years ago, the formation time of Nusplingen platy limestone had been calculated using comparisons to the deposition period of limestone in modern bodies of water. Now it has turned out, however, that Nusplingen platy limestone was built up predominantly by the carbonate exo-skeleton bearing gold algae, which is still in existence today. If the Emiliania Huxleyi which is living today is supplied with sufficient nutrients, it can, in only ten days, produce 0.5 to 1 cm of limestone sediment. More recent findings concerning micro evolutionary speciation show, moreover, that the biodiversity of the fossil sea creatures in Nusplingen platy limestone could have come about in the course of a few decades.

## Rapid calcification

Nusplingen platy limestone consists predominantly of gold algae exo-skeletons (Coccolithophorids). These minute algae (nanoplankton) floating in the sea shed a scaffold in the form of an annular carbonate shell-cover (coccolith).

The gold algae formed the food basis for the floating small sea lily (Saccacoma). More or less decomposed small sea lilies are finely dispersed throughout Nusplingen platy limestone, while it remains form a major component of thick shelves.

Small sea lilies could, given a large enough food supply, multiply on a massive scale and serve as a food supply for several ammonite genera. These three life forms (gold algae, small sea lilies and ammonites) formed an important food chain. Gold algae and small sea lilies occurred in great numbers and became rock formers, while the ammonites constituted the most common invertebrate fossils in Nusplingen platy limestone by far (1).

The pace of sedimentation of Nusplingen platy limestone was so rapid that belemnites were embedded obliquely or even vertically. In several shale levels, dead fish, which were also buried very rapidly, are embedded. This happened so fast that they did not have time to decompose.

## Gigantic algae blooms

In the summertime in cooler sea regions, today's gold algae produce so-called algal blooms, which can cover sea areas of up to 100,000 $km^2$. We talk of an algal bloom at concentrations of more than 1,000 cells per millilitre of water. Under these conditions, the algae double in number every 8.5 hours. In extreme cases, such an algal bloom can cover an area the size of England and produce a good 100 tons of lime.

## Rapid species formation

The only organisms in the Nusplingen platy limestone to consistently change through various layer sequences are the ammonites. Allegedly, in the past, the micro evolutionary development of ammonites from layer to layer is supposed to have proceeded substantially more slowly than is familiar to us from species living today.

The studies carried out by the biologist David Reznick and his team on small fish (Guppies – Poecilia reticulata) from predator-rich and predator-poor waters show selectively-generated shape changes after only eighteen generations (2). They thereby developed up to ten million times faster than suggested by the fossil sequence.

Rapid species formation was also observed as taking place under conditions of enormous environmental stress. For instance, in plants growing on mining waste heaps, contaminated with heavy metal (3), or mice exposed to environmental toxins (4).

further example of micro evolutionary development is the species-rich cichlid faunas in Lake Malawi, which came into existence in the last 200 years. Disturbed environmental conditions such as the lake's documented drying out phases have contributed to this, whereby, under different selection pressures, with basic forms with a very diverse genotype (genetic Polyvalence), new founder populations constantly came into being (5).

# References

1. Manfred Stephan. "Zur Bildungsdauer des Nusplinger Plattenkalks." *Studium Integrale* (April 2003): 12–20. http://www.wort-und-wissen.de/index2.php?artikel=sij/sij101/sij101-2.html.
2. David N. Reznick, Frank H. Shaw, F. Helen Rodd und Ruth G. Shaw. "Evaluation of the rate of evolution in natural populations of guppies (Poecilia reticulata)." *Science* 275 (28 March 1997): 1934–1937.
3. Reinhardt Junker. Prozesse der Artbildung, in S. Scherer (HG)'s Buch „Typen des Lebens", (Berlin, 1993): 31–45.
4. Silvia Garagna, Maurizio Zuccotti, Carlo Alberto Redi und Ernesto Capanna. "Trapping speciation." *Nature* 390 (20 November 1997): 241–242.
5. J. Fehrer. "Explosive Artbildung bei Buntbarschen der ostafrikanischen Seen." *Studium Integrale* (1997): 51–55.

# 31

# RAPIDLY RISING GRANITE DIAPIRS

Until a short time ago, most geologists were convinced that granite magmas move only very slowly in the form of rising diapirs from the lower crust to its final position location in the granite stock (Pluton). More recent observations of the rock composition and structure, laboratory measurements of the Earth's crust as well as fluid dynamic calculations show that the magmas in most cases flow upwards up to 100,000 times faster than previously thought. It therefore speaks for itself that many diapirs, which to date have been ascribed an age of millions of years, are, in reality, very young.

Granite is a fine-to-coarse textured crystalline rock, mostly light in color, with a high silcon content.

A diapir is the general term for a concentration material of low viscosity, circular at the base and mushroom-shaped at the top, which, due to upwardly pushing forces, rises through its high viscous surroundings. In addition to granite diapirs, we also talk of salt diapirs.

## The formation of granite

Hot magma rises to within a few kilometres below Earth's surface and usually forms an irregularly-shaped body of granite, also called a Pluton. Certain minerals already crystallise during this activity. However, the largest proportion of the mixture crystallises at the place of the intrusion during cooling.

A batholith is formed if, over the course of time, several granite stocks accumulate in a small area.

## Processes with nongeologically high progress:

Calculations have shown that an average molten mass can be transported in forty-one days through a 6 m wide and 30 km long Dike*. Thus, a batholith of 6,000 $km^3$ can form within only 350 years. A bit by bit filling over tens of thousands of years is impossible because the traces of such a process are missing. At the interface between old and young granite stocks, the old "cooled" stock would be reheated and recrystallised by the newly arriving hot one. Thickness and signs of reheating in the surrounding rocks of feeder-dikes** confirm this conclusion.

In certain cases, chemical analyses show that no chemical balance could be achieved between the molten mass and the remaining rock in the source area before the magma was withdrawn. If, in a short amount of time, a lot of magma formed in a narrowly delimited area under the Earth's crust, and the material experienced chemical homogenisation either before segregation or at the time of intrusion, these observations make sense.

## Epidote

A very strong indication for fast transportation is the mineral epidote, which is found in some batholiths. Epidote is only stable when it is about 20 km below the Earth's surface and in contact with magma. According to experiments, 0.5 mm grains of epidote from the Front Range disintegrate when they make their way into the upper crust in fifty years at 800 °C. In the case of White-Creek-Batholith, a flow rate of at least 700 m a year has been calculated based on the size of the grains found at the assumed temperature and depth, before the ascent of the magma. Thus, the creation of a batholith in decades to centuries is definitely realistic.

## Fast intrusion of granite molten mass through dikes

The controversy concerning magma transportation is in full swing (1). In spite of many gaps in knowledge, it can be established with remarkable clarity, that within the Earth's interior, large-scale processes are taking place, or at least have occurred for certain periods of the Earth's history, which are many magnitudes faster than the typically-estimated geological rates, such as the movement of plates in plate tectonics (presently several centimetres per year).

* Dikes are very extensive bodies of plate-like magmatic rock which fill in the larger fissures and cut or traverse the surrounding rock.
** Dikes, which act as feed channels for plutons, have been given the name "feeder-dikes."

## Reference

1. Franz Egli-Arm. *Studium Integrale* (April 1998): 6–16. http://www.wort-und-wissen.de/index2.php?artikel=sij/sij51/sij51-2.html.

# CHEMICAL EVOLUTION:

Chemical evolution is concerned with the genesis of basic building blocks for vitally essential molecules (amino acids), and with the development of biologically active molecules from these building blocks (proteins). It is not known how these proteins could have combined to form living cells.

## The primordial soup theory

According to conventional doctrine, the initial cells (the first form of life) supposedly developed spontaneously in a liquid known as primordial soup, in which all chemical substances required for its development were present. In the field of evolutionary biology, scientists make no attempt to explain the various primordial soup theories. They argue that the development of the first cell has nothing to do with evolution. This is the reason the title of this work *Ninety-five Theses Against Evolution* was chosen.

## Reproducible results

Various prebiotic conditions can be simulated in the laboratory and associated reactions studied. Any chemist can reproduce the results. The results concerning the primordial soup theory are extremely disillusioning. Among others, it has been observed that in an aqueous primordial soup and/or atmosphere containing oxygen, long chains of molecules never form. The water decomposes the molecules immediately and as soon as they encounter oxygen, they oxidize.

## 32

## VIVUM EX VIVO

"Vivum ex vivo" – Life comes only [sic] from life. This statement formulated by Louis Pasteur still complies fully even today with all data obtained experimentally from inanimate, prebiotic nature. At the time of Darwin, it was still believed that life could develop spontaneously in/through waste or rotting garbage. This development was called "abiogenesis." Louis Pasteur was the first to prove that bacteria cannot originate on its own.

On April 1, 1864, at the Sorbonne in Paris, Louis Pasteur proved before a large group of scientists that abiogenesis does not function. Pasteur, who rejected Darwin's theory of the origin of the species concluded that life can only evolve from life. Nevertheless, today many scientists still believe that several billion years ago on the primordial Earth, abiogenesis could possibly have resulted in so-called simple forms of life (1).

It is helpful to remember that even the simplest protozoa are comparable with a personal computer in terms of their complexity. Hundreds of mechanisms and hundreds of thousands of correct links are required for a cell to live. The failure of even one single mechanism (or its absence in a fully functional form from the beginning) results in the death of the cell or in it not being capable of living from the very beginning.

Nobel Prize winner Francis Crick saw the obvious impossibility of life evolving by chance. As an atheist and evolutionist, however, he did not want to accept a creator as the originator of life and therefore espoused the theory that life on

Earth originated from extraterrestrial sources. However, this does not solve the problem; it simply shifts it into space.

Renowned evolutionist and senior writer of *Scientific American* John Horgan wrote the following: "If I were a creationist, I would cease attacking the theory of evolution — which is so well supported by the fossil record — and focus instead on the origin of life. This is by far the weakest strut of the chassis of modern biology. The origin of life is a science writer's dream. It abounds with exotic scientists and exotic theories, which are never entirely abandoned or accepted, but merely go in and out of fashion"(2).

At a presentation in CERN near Geneva (17 Nov. 1965) the biochemist Ernest Kahane stated: "It is absurd and absolutely nonsense to believe that a living cell developed all by itself; nevertheless, I do believe it, because I cannot imagine any alternative."

## References

1. Bruno Vollmert. Das Molekül und das Leben, Rowohlt (1985, Der Urey-Miller-Versuch: Ursuppen): 39–45.
2. John Horgan. *The End of Science: Facing the Limits of Knowledge in the Twilight of the Scientific Age.* (London: Little, Brown & Co, 1997): 138.

# 33

# THE MILLER EXPERIMENT

In 1953, biologist and chemist Stanley L. Miller constructed a test apparatus to simulate the development of amino acids under primordial soup conditions. He was successful in synthesizing various simple amino acids by means of a spark discharge in a gas mixture over a number of days. However, his experiment has become insignificant in present molecular biology for a number of reasons. The water in Miller's primordial soup prevented the formation of chain molecules. Toxic substances also developed in his primordial soup. The experiment was accomplished under the exclusion of oxygen and otherwise does not coincide with today's knowledge regarding the primordial atmosphere of the Earth.

For some time, Stanley L. Miller's experiment was celebrated as a resounding success for evolutional theory. Even today, many school books still contend that life began with lightning acting on a particular primordial atmosphere. At that time, the first basic building blocks of life, the amino acids, supposedly evolved. The Miller experiment allegedly provides proof that such a scenario in natural science (1) is reproducible.

However, the primordial soup experiments performed by Stanley Miller and others have become uninteresting, because only fractions of the necessary basic building blocks for life develop under the described primordial soup conditions (2).

Biochemist Klaus Dose came to a sobering conclusion, still valid today, following the Eighth International Conference on

The Origin of Life. He had to recognize that "a major portion of the reaction products from simulation experiments were no closer to life than the contents of coal tar" (3).

In spite of, or perhaps due to, numerous experiments on the origin of life, the insight is increasing that natural chemical processes are not able to initiate life.

## Concentration and practical combination of amino acids

Even if amino acids were formed in a primordial soup, it would be necessary for them to concentrate in a further step and combine spontaneously to form practical molecular chains capable of carrying information. At least 500,000 base pairs are required to produce even the DNA of the simplest bacteria. Cell tissue, cell walls and various mechanisms must be formed simultaneously in order for the cell to be viable from the very beginning.

## References

1. Stanley L. Miller. "A Production of Amino Acids Under Possible Primitive Earth Conditions." *Science* 117 (15 May 1953): 528–529.
2. Paul Lüth. Der Mensch ist kein Zufall, DVA, (Stuttgart, 1981): 46–64.
3. Bruno Vollmert. Das Molekül und das Leben, Rowohlt (1985): 39–45.

# 34

# DNA (DEOXYRIBONUCLEIC ACID)

In 1953, Francis Crick and James Watson were successful in deriving a model of the three-dimensional structure of genetic material on the basis of X-ray diffraction data from Rosalind Franklin and Maurice Wilkins. Since this discovery, it has been clear that life is not possible without DNA. On the other hand, it also means that if these proteins did not exist previously, which is generally assumed today, the DNA must have developed by chance under primordial soup conditions without an auxiliary matrix. Laboratory experiments show sobering results. Such a scenario is not possible.

Chemically, DNA is nothing more than an extremely long molecular chain (polymer) consisting of three different types of building blocks: sugar molecules, nitrogen bases (nucleo bases) and phosphorus. Various problems already exist for the creation of these building blocks. DNA contains a specific type of sugar called D-ribose. The main problems with the chemical development of this compound are its short half life, only forty-four years, which is much too short by geological standards, and its unique three-dimensional structure.

It has been possible to produce two of the four essential nitrogen bases in primordial soup experiments with very low yield (0.5% for adenine and 0.1% for guanine). Cytosin and uracil were not available. Moreover, their life expectancy is also much too short. Even if the history of the genesis of life were to stretch over many billions of years, it is unimaginable that the substances required to build DNA would ever converge simultaneously and in purified form.

A further constituent of the DNA structure is phosphoric acid. Although phosphorus is present on the Earth, it is available only in the form of minerals with extremely low solubility (apatite and phosphorite). In this form, it would be impossible for phosphorus to contribute to the formation of a DNA chain molecule.

## Conclusion

Neither a sugar molecule, the four nitrogen bases nor phosphoric acid could evolve by themselves under natural conditions in forms practical for production of DNA (1). The associated question regarding the possibility of the synthesis of a DNA chain molecule from these molecules is therefore completely superfluous. Moreover, experience from polymer chemistry shows that it is not possible for DNA to evolve spontaneously without an auxiliary matrix (as is offered by a cell), even when all building blocks are present.

## Reference

1. Junker und Scherer. Evolution, ein kritisches Lehrbuch, Weyel (2006): 104–114.

# 35

# POLYMER CHEMISTRY

Today it is possible for chemists to produce amino acids with technical help. However, the formation of long chains such as required for living creatures is possible only under extremely clean conditions. Even minute impurities can lead to interruption of the chain. Moreover, the amino acid chains decompose as soon as they encounter water. Since such a hypothetic primordial soup would certainly have contained water, so it is impossible for such amino acid chains and particularly complete proteins to have formed.

A major problem for the origin of life is the fact that proteins decompose because of chemical laws as soon as they come into contact with water (1) (2). To make matters worse, water is produced during the production of the proteins. This process, called hydrolysis, disrupts polycondensation and immediately destroys any polymers developing. In living cells, a finely balanced process drains the water produced during protein production by special enzymes.

## Chain formation with bifunctional molecules

Molecules must be at least bifunctional in order to join (i.e., they must have two linkage points.) If a monofunctional molecule (i.e., a molecule with only one linkage point) attaches itself to the end of the chain, it is not possible for any further molecules to dock on, so chain formation is discontinued (3). Now, it is necessary to imagine that a primordial soup is not a polymer chemical laboratory where the various processes are supervised and the formation of chains is terminated specifically by adding monofunctional molecules only when the desired chain length has been reached (4).

The only environment known in which DNA strings form are the living cells. The prerequisite for creation of proteins is living cells which, for their part, also consist of proteins.

Without proteins there are no cells and without cells no proteins. Vivum ex vivo – Life comes from life only. This principle is again confirmed.

## References

1. J. Sarfati. "Origin of life: the polymerization problem." *Journal of Creation* 12 (1998): 281–284.
2. G.B. Johnson und P.H. Raven. *Biology, Principles & Explorations.* Florida: Holt, Reinhart and Winston, 1998: 235.
3. Bruno Vollmert. Das Molekül und das Leben, Rowohlt (1985): 54-58.
4. P.H. Raven. *Biology, A current bubble hypothesis.* (WCB/McGraw-Hill, 1999): 69.

# 36

# CHIRALITY

When amino acids and sugars, the most important building blocks for life, are produced in a laboratory, an equal number of left-turning and right-turning molecules is produced. However, virtually only left-turning molecules can be used to form living cells. Right-turning molecules have a toxic effect on the cell. Since the genome of the simplest known form of life already consists of approximately a half million building blocks, it is not imaginable that a sufficient number of primordial substances would ever occur under natural circumstances.

The basic building blocks for the genetic material as well as those for the proteins are characterized by the fact that their image cannot be superimposed on their mirror image in the same manner as the right and left hand. This property is known as chirality.

Production of chiral molecules is usually very complicated and always requires the presence of chiral information (e.g., from a chiral catalyst). Left-turning and right-turning molecules occur in equal quantities in all chemical processes producing amino acids and sugars.

Before life originated, a decision must have been made at some time in favour of right-turning or left-turning molecules. At the level of living organisms, it is imaginable that competition between individuals and species could have led to selection processes. A selection process based on competition is not imaginable for a mixture of substances in which both constituents contain the same chemical energy (1).

## Amino acids from space

Chemist Ronald Breslow from Columbia University in New York found an explanation for how exclusively left-turning amino acids could congregate in the model used for the primordial soup theory. At a convention of the American Chemical Society in New Orleans, he reported that amino acids which were brought to Earth by meteorites were subjected to radiation which had a higher tendency to destroy right-turning amino acids. This would result in an excess of left-turning amino acids.

Breslow and his colleagues were able to simulate in the laboratory how left-turning and right-turning amino acids combine with one another during crystallization so that only those amino acids that do not have a partner remain dissolved in the water. If the comet transported more left-rotating amino acids than right-rotating, it would be possible for a solution to develop in such a process, which contained almost exclusively left-turning amino acids (2). However, it can also be assumed that a slight excess of left-turning or right-turning amino acids would quickly equalise in a geological environment (3).

## Bacteria from right-turning amino acids

Since there are only few bacteria on Earth which consist of right-turning amino acids, at least two different comets must have fallen on the Earth. This would have to develop all twenty different amino acids essential for formation of a living cell twice.

# References

1. Junker und Scherer. Evolution, ein kritisches Lehrbuch, Weyel (2006): 108.
2. Spiegel Online. Chemiker simulieren Siegeszug linksdrehender Moleküle. (7 April 2008). http://www.spiegel.de/wissenschaft/mensch/0,1518,545766,00.html.
3. K. Dose. Präbiotische Evolution und der Ursprung des Lebens, Chemie unserer Zeit 21 (1987) :177–185.

# 37

# FOLDING OF PROTEINS

A protein must have a very specific three-dimensional shape in order to correctly exercise its function in a cell. Although only fractions of a second are required for a single protein folding to be completed in a cell, it would take billions of years to run through all of the possibilities for a single folding! Here it is necessary to know that an incorrectly-folded protein usually has a negative effect for the living creature (in the worst case, even death). For the formation of one single cell, it is necessary for thousands of proteins to be correctly folded. This leaves very little leeway for random processes.

The first step in the formation of protein is synthesis of a linear sequence of amino acids (primary structure). However, a protein can accomplish its actual function only when a precisely defined three-dimensional structure is present in addition to the specific sequence. This structure consists of characteristic structural elements (secondary structures), which for their part are folded in a higher-level spatial arrangement (tertiary structure). Moreover, aggregations of a number of proteins are known, which for their part have a specified structure (quaternary structure).

## The problem of protein folding

Proteins control nearly all functions in the human body. The folding determines the function of the proteins. Any change in the protein folding results in a change of function. Even a minute change in the folding process of an otherwise beneficial protein can cause disease.

Since the number of possible protein folding increases exponentially with the length of the amino acid chain, the time required to try all possible foldings (conformations) requires several billion years even for a small protein. In practice, however, a precisely-defined spatial structure is made within fractions of a second.

This phenomenon, known as the Levinthal paradox, clearly indicates that proteins obviously do not run through all possible foldings, but rather find short cuts on the way to their final structure with the aid of the so-called chaperones. The question which arises here is how the so-called chaperones know how the proteins have to look like in their final form. As with the formation of the primary structure, the formation of the tertiary and quaternary structure requires information, which cannot have originated on its own, because the final form must be known in advance.

## Virtual protein folding with Blue Gene

In 2005, IBM developed Blue Gene, the highest performance supercomputer in the world at that time in order to solve the problem of protein folding (1). The reason for this development is given on IBM's internet site: "The community of scientists considers the problem of protein folding to be one of the greatest challenges – a fundamental problem for science … whose solution can be achieved only by use of extremely high performance computer technology."

In spite of the enormous research power used here, it is estimated that Blue Gene would require approximately one year to complete the calculations for a model for folding a simple protein. A researcher at IBM mentioned, "The complexity of the problem and the simplicity with which it is solved in the human body is absolutely astounding" (2).

## References

1. IBM, Blue Gene Research Project, 2003. http://www.research.ibm.com/bluegene/index.html.
2. S. Lohr. "IBM plans supercomputer that works at the speed of life." *New York Times* (6 December 1999): C-1.

## 38

## ADDRESSING OF PROTEINS

A protein contains on average approximately 1,000 letters or amino acids. After a protein has been produced in a cell, it is necessary to transport it to the location where it is to be used. For this purpose, each individual protein is given a complex address. Random development or assignment of this address is not imaginable. Moreover, incorrectly addressed proteins are in many cases not only useless, they can even be damaging.

Proteins are not produced at the location they are finally needed. There are a great number of incorrect locations where a newly-formed protein could be transported, however only very few locations, and frequently only one single location, fulfils the correct purpose.

### But how do proteins find the correct destination?

Newly-formed chains of amino acids contain a section in which the address is given. This address is the location where they are to be used. Each section is normally attached to the end of the longer chain representing the protein. Every correctly folded protein will fit at a certain location and must be properly addressed accordingly. A protein at an incorrect location is, however, much more dangerous than an incorrectly delivered letter, because it can cause a disease (1).

For a cell to function, it is not only necessary to produce the correct proteins, but also to solve the complex problem of precise addressing (2). To clarify the problem, during every minute of our human existence, it is necessary to produce not only one or two, but rather millions of properly addressed proteins in our body which then must be transported and properly built in.

It is unrealistic to believe that such sequences could occur in an incremental process controlled by chance.

## References

1. John Travis. "Zip Code plan for proteins wins Nobel." *Science News* 156 (16 October 1999): 246.
2. Guenter Blobel. Britannica Biography Collection.

# 39

# PRODUCTION OF PROTEINS

It is not sufficient that all proteins produced in a living cell are folded correctly and addressed correctly. It is also necessary to produce the correct quantity of each protein. If it were not possible for a cell to stop production of a certain protein at the correct time, this would have an effect similar to slowly burning down the whole house instead of simply adding some wood to the fire in the fireplace. The mechanism that stops and starts production of proteins must be correctly functional in each cell from the very beginning.

It cannot be taken for granted that the production of each individual protein is started and stopped at the correct time (1). Life begins when the cell contains the correct quantity of each protein, when all proteins are correctly folded and integrated at the correct location.

However, as life begins, the proteins also begin to wear out. It is then necessary for the cell to be capable of replacing the worn-out proteins with newly-produced proteins. This mechanism must also be present and fully functioning from the very beginning.

## DNA regulation sequence and regulator proteins

The most important feature for starting and stopping protein production consists of the regulation sequences on the DNA. These sections of the DNA have the function of telling the cells when to start and stop producing the various proteins. However, the DNA itself cannot start or stop protein production. This requires cooperation with special regulator proteins that are folded in such a manner that they precisely match a special section of the DNA (2).

## Conclusion

The DNA regulation sequence and the regulating proteins require one another mutually. Both must be perfectly coordinated to correctly start and stop production of the associated proteins. Actually, they form a switch, an irreducible complex system that is essential for life.

## References

1. S. Aldridge. *The Thread of Life: The story of genes and genetic engineering.* (Cambridge: Cambridge University Press, 1996): 47–53.
2. B. Alberts, D. Bray und A. Johnson, et al. *Essential Cell Biology. An Introduction to the Molecular Biology of the Cell.* (New York: Garland Publishing Inc., 1998): 259–262.

# 40

# INTERNAL CELL CONTROL MECHANISMS

Internal cell control mechanisms ensure that faulty proteins are broken down into their constituents and used elsewhere. These devices would counteract any macro evolutionary development, because they also eliminate proteins which would provide a benefit for the organism, if they do not fit into the existing concept. Even the DNA string is checked and corrected continuously during the copying operation at cell division. Life appears to be established to the principle of preserving the existing proteins.

Since defective proteins can have a damaging effect to living cells and would also use up unnecessary resources, incorrectly formed proteins are broken down immediately. This system eliminates many if not most proteins which have mutated by chance (1) (2).

## Consequences for evolution theory

According to the doctrine of the evolution theory popular today, mutations only are capable of producing new genetic information. The known control mechanisms acting inside the cell therefore pose a major obstacle, from an evolutionary point of view, for the assumed development of life.

Assuming that the various forms of life on our planet were created "each after their kind (species)" (3), these control mechanisms to maintain the individually matched basic structures have a clear objective.

# References

1. C. Lee und M.H. Yu. "Protein folding and diseases." *Journal of Biochemistry and Molecular Biology* 38 (2005): 275–280.
2. Walid A. Houry, Dmitrij Frishman, Christoph Eckerskorn, Friedrich Lottspeich und F. Ulrich Hartl. "Identification of in vivo substrates of the chaperon in GroEL." *Nature* 402 (11 November 1999): 147–148.
3. The Bible, Gen. 1:11–14; 20–25.

# RADIOMETRY AND GEOPHYSICS

Radiometric measuring methods are frequently used for determining the age of rock formations or organic samples. This method utilizes the fact that some materials contain or could have contained unstable radioactive isotopes.

Some isotopes are unstable or radioactive, meaning that they decay, sooner or later forming other daughter isotopes. It is possible to measure the relationship between the daughter and parent isotope. This relationship allows us to draw conclusions regarding the radiometric age of the specimen in question based on the half-life* of the parent isotope. Moreover, the radiation may cause during such decay processes some visible radiation damage (radiation halos and/or fission tracks), which can also be age-interpreted.

It is necessary to make three assumptions if radiometric measurements are to make sense:

a) The half-life must remain constant over the entire decay period.
b) No parent or daughter isotopes should have left the probe or been added.
c) The initial conditions of the probe must be known.

Based on the initial conditions assumed under (c), it is possible to calculate a radiometric age of many million years for most geological strata. However, systematic deviations occur when the same material is analyzed with different methods. Some findings indicate that accelerated radioactive decay may have occurred some times on our Earth. So (a) is not satisfied.

Since the majority of non-radiometric age determination methods in geology, paleontology and geophysics reveal ages which are more recent by several orders of magnitude, it is necessary to consider critically the radiometric data. The magnitude of erroneous estimates in this sector of science is shown by the lava on Hawaii which can be proven to be 200 years old; however, it is radio dated at several million years (1) (2).

* The half-life of a certain radioactive material is the period of time to decay to half of its initial value:

## References

1. G.B. Dalrymple. *The Age of the Earth.* (Stanford University Press, 1991): 91.
2. Andrew A. Snelling. "Excess Argon: The Achilles' Heel of Potassium-Argon and Argon-Argon Dating of Volcanic Rocks." Institute for Creation Research, 1999. http://www.icr.org/article/excess-argon-achillies-heel-potassium-argon-dating (Note: Although the spelling is "achilles" and not "achillies," the incorrect spelling used in the internet address was not corrected to make it possible to call the website in question).

# 41

# DEVIATIONS IN RADIOMETRY

Various radiometric methods can be used to determine the age of a rock, depending on whether the rock contains various unstable radioactive isotopes. Generally, for reasons of cost, only one single method is used. However, if the same rock is measured using different methods, it is possible to get distinguished and systematic deviations.

Today, a number of various methods based on radioisotopes are used for determination of age. If the results of these methods are to be credible, they should agree at least within the usually estimated tolerance limits. As a rule, they do not even come close. Since the measurements show systematic and repeatable deviations, a systematic error must be present in the measuring methods and/or the evaluation.

For confirmation of such observations, a piece of Cardenas basalt, a lava stone from the Grand Canyon with a conventional age of allegedly 1.1 billion years, was analyzed using four different methods (1). Here are the results:

| | |
|---|---|
| Potassium-Argon: (beta decay) | 516 million years (tolerance +/- 30 million) on 14 specimens |
| Rubidium-Strontium: (beta decay) | 892 million years (tolerance +/- 82 million) on 22 specimens |
| Samarium-Neodym: (alpha decay) | 1588 million years (tolerance +/- 170 million) on 8 specimens |
| Lead-Lead: (alpha decay) | 1385 million years (tolerance +/- 950 million) on 4 specimens |

These studies showed that specimens subject to alpha decay usually show higher age values than those subject to beta decay. During alpha decay, helium nuclei are formed while beta decay radiates electrons. This shows that the apparent age is higher the heavier the atoms of the parent isotope.

Unfortunately, only few comparative measurements have been performed to date. For this reason, the statistical relevance is relatively low. On further specimens from ten different locations, the measured results were so different that evaluation was not possible. Others, by contrast, could be evaluated well, however yielded distinctive and systematic deviations (2).

## Accelerated radioactive decay

A possible explanation for the systematic differences is that the radioactive decay was accelerated during a certain period. It is imaginable that the Earth's crust was subject to massive neutron radiation during its early development and/or a catastrophic event for a limited time with an increased production of daughter isotopes.

## Conclusion

It would be desirable for public universities to perform comparative measurements to an increasing extent on materials on which various methods could be used. Since this has hardly been done to date, it is difficult to avoid the suspicion. Someone does not necessarily want the results of radiometry to be questioned. When a rock is tested for its age, normally only one of the possible methods is used.

## References

1. Don DeYoung. *Thousands... not Billions, Challenging an Icon of Evolution*. (Master Books, 2005): 126.
2. Larry Vardiman, Andrew A. Snelling, Eugene F. Chaffin. *Radioisotopes and the age of the Earth*, Vol. 2. Institute for Creation Research, El Cajon, CA, 2005): 422.

## 42

## ACCELERATOR MASS SPECTROMETER (AMS)

With a state-of-the-art accelerator mass spectrometer (AMS) it should be possible to analyze up to 90,000 years old carbon-containing material (graphite, marble, anthracite and diamonds). However, to date, not one single type of material has been found with a radiometric age to exceed 71,000 years. These age figures, which are far too low for the conventional doctrine, are explained with contamination. However, great efforts have not been able to prove such contamination. Moreover, it is imaginable that less radioactive carbon (C-14) was present in the primordial Earth atmosphere. If so, the materials studied would have to be classified as even younger.

The carbon isotope C-14 has a half-life of 5,730 years. It decays to form nitrogen. Conclusions regarding the age of a specimen can be made by measuring the ratio of C-14 to C-12 in a material containing carbon. Specimens older than 90,000 years will no longer contain any measurable quantity of C-14. Nevertheless, various carbon specimens that are allegedly between 34 and 311 million years old (1) still contain 0.1 to 0.46% C-14. This corresponds to a maximum radiometric age of 57,000 years.

If the Earth's magnetic field was stronger in the earlier days of the Earth (which can be assumed) (2), even these 57,000 years are estimated too high. A stronger global magnetic field would effectively reduce the cosmic radiation thereby producing less C-14. It can therefore be assumed that at the beginning, less C-14 was contained in the specimens.

## Tests on diamonds

Diamonds are particularly interesting for such tests. Astrophysicist Larry Vardiman and his team tested twelve different diamonds originating from five different locations. The average C-14 content was 0.09% corresponding to a maximum age of 58,000 years (3). According to conventional geology, however, the diamonds would have had to be up to three billion years old. However, if they were even approximately this old, they should no longer contain any traces of C-14. The objection that the specimens were contaminated in the course of time hardly applies for diamonds. According to present knowledge, diamonds cannot be contaminated (4).

## References

1. Larry Vardiman, Andrew A. Snelling, Eugene F. Chaffin. *Radioisotopes and the age of the Earth*, Vol. 2. (Institute for Creation Research, El Cajon, CA, 2005): 605–606.
2. Russel Humphreys. "The Earth's magnetic Field is young." *impact* 242 (August 1993).
3. Don DeYoung. *Thousands ... not Billions, Challenging an Icon of Evolution.* (Master Books, 2005): 46–62.
4. Vardiman, 609

# 43

# URANIUM, HELIUM AND LEAD IN ZIRCONIUM

Zirconium crystals are found in granite around the world. Some of these crystals also contain small quantities of uranium, which is subject to radioactive decay producing helium and lead, both stable substances. It is possible to calculate the age of the crystals based on the quantity of helium present today and the rate at which it continuously escapes, commonly called its diffusion rate. Interestingly enough, rock strata that are allegedly billions of years old, occasionally contain zirconium that is only 4,000 to 8,000 years old on the basis of helium content.

The uranium frequently contained in zirconium crystals is an unstable element, which decays to form lead and helium in the course of time. While the small helium atoms slowly escape from the crystal after decay, the larger uranium and lead atoms remain trapped in the crystal. The more uranium decays to form lead and helium, the older the crystal must be. It is possible to draw conclusions regarding the age of the crystal by comparing the ratio of uranium atoms to lead atoms. This quite frequently results in age figures of over one billion years.

## Helium diffusion rate

However, if the quantity of helium still present in the crystal is taken into consideration, an age of 4,000 to 8,000 years (1) can be calculated, considering the quantity of helium escaping from the crystal per time unit (helium diffusion rate).

If the uranium had decayed at the same rate as today in the course of a billion years, the helium would have been able to escape from the crystal continuously over this long period and we would hardly find any helium still present in the crystal today (2). The quantity of the helium content is therefore an indicator that the crystals were occasionally subjected to strong radiation leading to accelerated decay (3).

## Legitimate criticism

Unfortunately, to date it has only been possible to use specimens from one single deep well, making the worldwide significance of these results dubious. Moreover, the location selected for the well was in the vicinity of natural helium deposits which could have lead to contamination. In any case, it would be desirable for public universities to examine this subject.

## References

1. Larry Vardiman, Andrew A. Snelling, Eugene F. Chaffin. *Radioisotopes and the age of the Earth* 2. (Institute for Creation Research, El Cajon, CA, 2005): 56.
2. ICR Acts & Facts, Vol. 31. No. 10. (October 2002).
3. Don DeYoung. *Thousands ... not Billions, Challenging an Icon of Evolution.* (Master Books, 2005).

# 44

# RADIOACTIVE DECAY TO LEAD

Frequently, the quantity of uranium 238 and lead 206 are measured for radiometric determination of the age of rocks. The half-life with which uranium 238 decays to form lead 206 is 4.46 billion years. After 4.5 billion years (the alleged age of the Earth), at least the same quantity of lead and uranium should therefore be present on the Earth's surface. However, in reality, there is more lead than uranium. It could be assumed that an undetermined quantity of lead 206 formed directly when the rocks originated. Moreover, in addition to uranium 238, 52 other elements also decay to form lead 206. The half-life of these elements varies between a few microseconds and 245,500 years. It is therefore not possible to estimate how much of the lead 206 present today actually originated from uranium 238.

Various methods are used for radiometric determination of the age of rocks. The principle is always the same. An unstable or radioactive initial material decays to form a different stable element. The subsequent list shows how many other unstable elements also decay to form the same stable element:

Potassium – Argon → 3 other elements also decay to Argon
Rubidium – Strontium → 4 " " Strontium
Samarium – Neodym → 13 " " Neodym
Lutetium – Hafnium → 10 " " Hafnium

| | | | | |
|---|---|---|---|---|
| Rhenium – Osmium | → | 9 " " | Osmium |
| Thorium-232 – Lead 208 | → | 26 " " | Lead 208 |
| Uranium 235 – Lead-207 | → | 45 " " | Lead207 |
| Uranium 238 – Lead 206 | → | 52 " " | Lead 206 |

Generally, only the decay of uranium 238 is taken into consideration when using the uranium 238/lead 206 method (1). All other elements, which also decay to form lead 206, are ignored.

## Decay sequence from uranium 238 to lead 206:

| | | |
|---|---|---|
| Uranium 238 | decays with a half-life of | 4.46 billion years to |
| Thorium-234 | " " | 24.1 days to |
| Protactinium-234 | " " | 46.69 hours to |
| Uranium 234 | " " | 245,500 years to |
| Thorium-230 | " " | 75,400 years to |
| Radium-226 | " " | 1,599 years to |
| Radon-222 | " " | 3.82 days to |
| Polonium-218 | " " | 3.04 minutes to |
| Lead 214 | " " | 27 minutes to |
| Bismuth 214 | " " | 19.9 minutes to |
| Polonium-210 | " " | 0.16 milliseconds to Lead 206. |

Lead 206 is stable.

Using the model of a young Earth, the origin of the radiogenic lead present today could be traced back to several short-lived isotopes. In the model with a 4.5 billion-year-old Earth, the radiogenic lead can be traced back exclusively to the decay of long-lived isotopes. Both interpretations are speculative.

The decay time of short-lived isotopes is shown on the following internet site: http://nucleardata.nuclear.lu.se/nucleardata/toi/sumframe.htm. (Specify "atom mass" and click on "show drawing.")

# Reference

1. Charles W. Lucas Jr. "Radiohalos – Key Evidence for Origin/Age of the Earth." Proceedings of the Cosmology Conference 2003, Ohio State University, Columbus, Ohio.

# 45

# RADIOACTIVE DECAY AT PLASMA TEMPERATURES

When known radioactive materials are heated up to plasma temperatures, the half-life of elements such as uranium 238 decreases from 4.5 billion years to 2.08 minutes. This circumstance represents another uncertain factor for radiometric age determination.

Plasma temperatures normally do not occur. Anyway, this demonstrates that radiometric decay times are not constant.

If a solid substance is heated, most elements first become liquid and then gaseous after reaching a certain temperature. If this gas is heated even further, it turns to plasma at extremely high temperatures. This plasma then has characteristics completely different from the gas from which it originated. Among other factors, the half-life of radioactive isotopes is reduced dramatically. The higher the temperature, the greater the reduction in half-life.

If the following materials are heated up to 15.4 billion degrees Kelvin, the half-life changes as follows (1) (2).

Uranium 238 decreases from 4.5 billion years to 2.08 minutes

Thorium-232 decreases from 14 billion years to 15.6 minutes

Samarium-147 decreases from 106 billion years to 1.56 minutes

Rubidium-87 decreases from 47 billion years to 2.46 minutes

Potassium-40 decreases from 1.2 billion years to 5.87 minutes

# References

1. Edward Boudraux. "Attenuation of accelerated decay rates by magnetic effects." Proceedings of the Cosmology Conference 2003, Ohio State University, Columbus, Ohio.
2. Edward Boudraux. "Accelerated Radioactive Decay Rates, a Minimal Quantitative Model." Proceedings of the Cosmology Conference 2003, Ohio State University, Columbus, Ohio.

# 46

# URANIUM AND POLONIUM RADIOHALOS

The frequency of uranium and polonium radiohalos in granite from the Palaeozoic and Mesozoic periods, allegedly 251 to 542 million years ago, indicates one or more phases of temporarily accelerated radioactive decay. Therefore, the results of radiometric measuring methods, including the fission track method, can be explained very well using the model of a young Earth.

Among other substances, granite contains biotite or black mica which contains a very small quantity of uranium. If this uranium is concentrated at spots, the decay of the uranium can form microscopically visible radiohalos (1).

Larry Vardiman and his team studied three groups of granite specimens (2):

- One from the Precambrian era (allegedly 542 million to 4.5 billion years ago)
- One from the Palaeozoic-Mesozoic period (allegedly 251 to 542 million years ago)
- One from the Cenozoic period (allegedly up to 250 million years ago)

It was noted that the frequency of radiohalos in the Palaeozoic era was occasionally significantly higher than in the other formations. This means that during this period, accelerated radioactive decay occurred with high probability. This breach of the rule makes a uniform interpretation of the development of these geological formations impossible. Why are significantly fewer radiohalos present in the upper and lower strata than in the middle strata? Especially in the allegedly four billion years long Precambrian period, significantly more radiohalos could be expected than in the Palaeozoic/Mesozoic period which lasted only a few hundred million years.

## Results of fission track method

So-called fission tracks result in zirconium crystals during radioactive decay of uranium. In this process, a number of atoms are knocked out of the normal crystal lattice leaving tiny tracks. After treatment with a suitable etching media, these tracks can be enlarged to make them visible under a microscope. The age of the crystal can then be calculated from the number of tracks and the heavy atoms that have not yet decayed.

The most frequent material to produce fission tracks is uranium 238. If it splits, it forms palladium 119 producing a fission track which can be observed in transparent material as well as in natural glass. When the specimen is temporarily heated by fifty to four hundred degrees, the tracks vanish. This means that all specimens containing fission tracks provide information of their thermal history. Determination of the age using the fission track method does not indicate the age of the rock, but primarily the time since the specimen was last heated to a significant extent.

The fission tracks can be counted after cleaning and etching the specimen. Then, the number of uranium 238 atoms which have not yet decayed are counted using a suitable measuring method. Conventional evaluation of the measured results, render an Earth history extending over millions and billions of years. However, if we consider the temporarily accelerated decay indicated by the uranium and polonium radiohalos, the results of the fission track method coincide quite well with the model of a young Earth (3).

## References

1. Robert V. Gentry. *Creation's Tiny Mystery*. (Earth Science Associates, May 1992): 214.
2. Larry Vardiman, Andrew A. Snelling, Eugene F. Chaffin. *Radioisotopes and the age of the Earth* 2 (Institute for Creation Research, El Cajon, CA, 2005): 101-207.
3. Don DeYoung. *Thousands ... not Billions, Challenging an Icon of Evolution*. (Master Books, 2005).

# 47

# HELIUM FROM INSIDE THE EARTH

Radioactive decay processes which occur inside of the Earth produce helium and heat. However, the quantity of helium escaping amounts to only four percent of that expected in relation to the escaping heat. One possible explanation is that the major portion of the helium is retained on the inside of the Earth. Another possibility would be that the Earth still has a large store of heat from its origin, meaning that not all heat results from radioactive decay. Neither of these possibilities is compatible with the model of an old Earth.

Helium escapes from the inside of the Earth because of radioactive decay. Simultaneously, heat is produced which escapes to the Earth's surface. The decay of uranium, thorium and potassium to produce the inert gases helium and argon is undisputed. The quantity of heat produced during this decay process should correspond to the quantity of helium escaping from the Earth' crust. To generate one Joule of heat it would be necessary to produce $10^{12}$ He atoms and $2 \times 10^{11}$ Ar atoms inside the Earth.

Now, the heat and helium escaping from the inside of the Earth has been measured:

The heat escaping through the Earth's crust in the area of the oceans is $0.1 W/m^2$. The quantity of helium escaping from the inside of the Earth in the area of the oceans is $4 \times 10^9$ helium atoms per square metre per second. The resulting calculated helium/heat ratio is $4 \times 10^{10}$ atoms per Joule, amounting to only four percent of what is expected on the basis of the quantity of helium escaping (1).

Two alternatives and two problems with the model of a 4.5 billion year old Earth:

a) Imagine that no helium at all is present on the inside of the Earth. If we now start a decay process producing the quantity of heat we measure today, it could, in fact, be expected that initially only a small portion of the helium would find its way to the Earth's surface. The major portion would be retained by the Earth's crust. However, a balance would have to build up in the long run. It is not imaginable that an incongruity between the quantity of helium produced and the quantity escaping still exists after 4.5 billion years.

b) Imagine that the quantity of heat escaping is the result of radioactive decay processes to only a small part. This is also hardly imaginable using a model of a 4.5 billion-year-old Earth. It is difficult to imagine that the Earth has still not cooled off after such a long period.

## Helium in the Earth's atmosphere

Interesting in this context is also the fact that the Earth's atmosphere contains less helium than predicted by the model of an old Earth. However, it is not yet clarified how much helium can escape from the atmosphere into space and how many helium nuclei are added to the Earth's atmosphere by solar wind (2).

# References

1. E. Ronald Oxburgh und R. Keith O′Nions. "Helium Loss, Tectonics, and the Terrestrial Heat Budget." *Science* 237 (25 September 1987): 1583–1588.
2. Melvin A. Cook. "Where is the Earth′s Radiogenic Helium?" *Nature* 179 (26 January 1957): 213.

# 48

# THE EARTH'S MAGNETIC FIELD

Most planets have their own magnetic field, also the sun. One would expect these magnetic fields to have a longer or shorter life depending on the hypotheses of the origin. Measurements of the Earth's magnetic field have indicated a continuous decrease over the last approximately 170 years. Based on these measurements, it is possible to estimate the age of the Earth's magnetic field to be fewer than 10,000 years old.

The Earth has a solid inner core of iron surrounded by a liquid core which is surrounded by a solid stone crust. The globe rotates causing the liquid portion of the core to move in a helical form due to Coriolis force. One can imagine that this motion has generated a dynamo that could have been responsible for development of the initially weak magnetic field of the Earth (dynamo theory). However, it has not yet been possible to reproduce this with any satisfactory mathematical model. On the contrary, the measured data indicates that the Earth was created with a relatively strong magnetic field, which continuously decreases (1).

The Earth's magnetic field has been measured since 1835. The measurements show that the field strength decreased by eight percent between 1835 and 1965. The various measurements allowed the conclusion that the magnetic field's strength may be reduced by half every 1,465 years. Measurements performed by the geophysical observatory in Munich show that the Earth's magnetic field has decreased for approximately 3,000 years. If it continues at this rate, it will no longer exist in 4,000 years (2).

## Reversal of the poles of the Earth's magnetic field

During development of geological strata, all magnetisable particles are aligned according to the current effective magnetic field and fixed in this direction. It has been found that the Earth's magnetic field has reversed its polarity many times in the past.

According to common doctrine, the polarity reversed every 250,000 years on average. This number is obtained by comparing the polarity reversal events documented in the geological strata with radiometric age. However, an angular change of approximately six degrees per day was observed in lava flows on Steen's Mountain, Oregon, USA (3). This means the local magnetic field could have reversed its polarity within approximately thirty days at the time of lava flow (4).

The present decrease of the Earth's magnetic field could initiate a reversal of the polarity. But this is unlikely. Such an event would happen fast. It is possible that the magnetic field reversed several times during the great flood and that its intensity is decreasing since creation.

## The common dynamo theory

It is not possible to answer many important questions with the common dynamo theory (5). This applies particularly to the question of how the gigantic quantities of liquid iron inside the Earth could have changed the direction of the magnetic flux. Was the Earth's magnetic field actually created by the rotation of the iron core? The only fact that can be assumed with high probability is that the position of the poles has changed only insignificantly during the Earth's history (6).

# References

1. D. Russel Humphreys. "The Earth's Magnetic Field is still losing energy." *CRSQ* 39/1 (March 2002): 3–13.
2. Geophysikalisches Observatorium in München, 3sat nano, (29 August 2002). http://www.3sat.de/nano/bstuecke/36057/index.html.
3. R.S. Coe, M. Prévot und P. Camps. "New Evidence for extraordinarily rapid change of the geomagnetic field during a reversal." *Nature* 374 (20 April 2002): 687–692.
4. R.S. Coe und M. Prévot. "Evidence suggesting extremely rapid field variation during a geomagnetic reversal." *Earth and Planetary, Science Letters* 92/3-4 (April 1989): 292–298.
5. M.R.E. Proctor und A.D. Gilbert. *Lectures on Solar and Planetary Dynamos*. Cambridge University Press, 1994.
6. Proceedings of the NATO Advanced Study Institute, "Theory of Solar and Planetary Dynamos," Isaac Newton Institute, Cambridge University, 20 Sept. to 2 Oct.1992.

# 49

# SALT MOUNTAINS AND SALT CONTENT OF OCEANS

In spite of extensive rainy periods (pluvial) during the quaternary era, which allegedly started 2.6 million years ago, the salt diapir Kuh e Namak in Central Iran was lifted more than 300 metres above the ground. If this salt mountain was as old as officially estimated, it should have been dissolved long ago. A further factor is that the salt from such salt domes is carried into the seas contributing to the slow increase in salt concentration of the oceans. If the import and export of salt into the world's oceans is measured, we conclude that the current process has been in progress for a maximum of sixty-two million years. This calculation is based on the unrealistic assumption that originally there was no salt in the world's oceans.

The most striking substance contained in seawater is a mixture of various salts. When the seawater evaporates, the salts are left. The evaporated water will form clouds. When these drift over the continents and cool down, rain falls. The rain water seeps into the ground, dissolving products of weathering such as lime and salts. Some of the water returns to the surface as spring water, flowing through streams, rivers and the ground water back into the sea (1).

The current salt content of the oceans as well as all exports and imports of salt can be measured today (2). It has been shown that the import of salt today is significantly higher than the export. If the current processes had continued for 3.5 billion years, the world's oceans would contain fifty-six times the salt content they have today (3).

## Rising salt diapirs

Much indicates that the Earth's climate during the Tertiary period, allegedly 2.6 to 65 million years ago, was significantly warmer worldwide and considerably rainier than in the tropics today. Even if a salt dome such as Kuh e Namak had risen as a mountain of salt in an occasionally desert dry climate, it could hardly have withstood a quaternary period lasting 2.6 million years without having been completely dissolved (4), since considerable rainy periods occurred during the Quaternary era.

Moreover, it can be concluded that the Quaternary rainy periods were considerably shorter than commonly assumed. It will probably be necessary to reduce the age of the geological formations around Kuh e Namak by several decades (5).

## References

1. E.K. Berner und R.A. Berner. *The global Water Cycle*. (Englewood Cliffs, NJ: Prentice-Hall, Inc., 1987).
2. Bryan Gregor, et al. "Chemical Cycles in the Evolution of the Earth," 1988.
3. Steven Austin und D. Russel Humphreys. "The sea's missing salt." Proceedings of the Second International Conference on Creationism, (1990):17–33.
4. Detlef Busche, Reza Sarvati und Ulf Siefker, Kuh-e-Namak: Reliefgeschichte eines Salzdoms im abflusslosen zentraliranischen Hochland, Petermanns Geographische Mitt. 146/2 (2002):68–77.
5. Manfred Stephan. "Langzeitproblem: Entstehung eines Salzbergs im Iran." *Studium Integrale* (April 2007): 12–20.

# 50

# NICKEL IN SEAWATER

Nickel ore is transported into the oceans by river water. Conclusions can be drawn regarding the age of the oceans based on the quantity of nickel added annually and the current nickel content of the seas. Here, we note that according to the present processes it would have taken a maximum of 300,000 years to achieve the present nickel content. Since no mechanism is known which removes nickel from seawater, it is not realistic to assume that our oceans are many millions of years old.

The following initial data is known:

a) On average, the rivers of the Earth carry 0.3 micrograms of nickel per litre of water into the sea (1).
b) The total quantity of water flowing into the sea from rivers and streams is 37,400 km³ annually on average.
c) The average nickel content of seawater is 1.7 micrograms per litre (2).
d) The quantity of water in the seas is $1.35 \times 10^{21}$ kg (3).
e) It is estimated that there are $2 \times 10^{14}$ kg of manganese nodules at the bottom of the Pacific Ocean, containing 0.63 percent nickel (4).

From these figures, it is possible to calculate the maximum time required to reach the current nickel contents based on the present processes. In order to calculate the highest age possible, it is assumed that there initially was no nickel present in ocean water or in the manganese nodules. Moreover, the interstellar dust from space, which also added nickel to the seas, is ignored.

Even under these circumstances, the maximum age that can be calculated for the Earth is only 300,000 years. Since no known mechanism can remove nickel from seawater, it is not possible to imagine a sea which is several million or even billion years old. If there was actually some sort of a worldwide flood which literally washed out the continents, it would be necessary to reduce again this figure of 300,000 years drastically.

Concerning the manganese nodules, the lime sludge which is deposited at the bottom of the oceans amounts to 1,000 to 10,000 times more than the manganese nodules. This means that the manganese nodules visible today should have been covered long ago, if the age calculated above is correct (5). The argument that the lime sludge covering the manganese nodules is continuously removed is hardly viable, because corresponding deposits cannot be found.

## References

1. W.H. Durum und J. Haffty. Geochimica et Cosmochimica Acta, 27 (1963): 2. D.A. Livingstone, Chemical composition of rivers and lakes, Geological Survey Professional Paper (1963): 48.
2. Hg. von J.P. Riley & G. Skirrow. Chemical Oceanography. (New York: Academic Press, Vol. 1, 1975) Auflage, 418.
3. Hg. von J.P. Riley & G. Skirrow, 2.
4. Eugen Seibold und Wolfgang H. Berger. *The sea floor*. (Springer Berlin, 1996): 289, 293.
5. Eugen Seibold und Wolfgang H. Berger, 291.

# 51

# PETROLEUM, COAL AND PETRIFIED WOOD

The claim that it requires long periods of time for the development of oil, coal or petrified wood is obsolete. Rapid formation of oil has already been tested experimentally for some time and in 2006 it was discovered that coal can form overnight under favourable conditions. Patents have already been registered for some years for petrifaction of wood. Petrified wood is used, for example, for tabletops and chimney plates.

## Formation of coal

An announcement by the Max-Planck Institute indicates that straw, wood, moist grass or leaves can be converted to coal overnight (1). A process has been presented with which plant biomass can be converted virtually completely to carbon and water directly without complicated intermediate steps. The process is called hydrothermal carbonization. It works like a steam pressure cooker, only at higher temperatures.

The cooking recipe for coal is amazingly simple. A pressure vessel is filled with any type of vegetable products, for example, with leaves, straw, grass, pieces of wood or pinecones. Then water and a small quantity of citric acid are added. The vessel is then sealed and the content is heated under pressure for twelve hours at 180 degrees Celsius. After the mixture has cooled down, the vessel is opened. It contains an aqueous black broth with finely distributed spherical-shaped particles of carbon called colloids.

All carbon previously contained in the vegetable material is now in the form of these particles as small, porous lignite balls.

## Rapid formation of coal in nature

In the Triassic or Jurassic period, carbon strata developed within only twenty-five to thirty years. This is revealed by the oval and circular concentric polonium radiohalos present in the material. Polonium-210 has a half-life of 138.4 days. If the radiohalos developed before compression of the coal stratum, they all would have to be oval (2).

## Formation of petroleum

For formation of sediment basins and the petroleum reserves present therein, geoscientists assume protracted processes lasting millions of years. By contrast, hydropyrolyses laboratory experiments on bedrock from sediment basins showed that petroleum can be formed and extracted very quickly at appropriately high temperatures or under suitable catalytic conditions (3).

As reported by geologists Borys M. Didyk and Bernd R.T. Simoneit, a 500 m thick deposit of phytoplankton, freely suspended seaweed from which petroleum escapes (4), is present in the Guaymas basin in the Gulf of California. Hypothermal fluids with a diameter of 8-12 cm are present at the surface of these sediments from which hot water exits at a temperature of 200°C. This water carries small spheres of oil with a diameter of 1-2 cm with it. Detailed research has shown that the composition of this oil is very similar to common petroleum. Age measurements using the radiocarbon method indicated 4,200-4,900 years. The oil formed at a temperature of over 315°C at a pressure of 200 bars. The estimate of the quantity of oil formed showed that use of the oil would be worthwhile if it could be collected.

## Petrification of wood

When wood is deposited in rivers, lakes or in the sea and is covered quickly enough with sediment, the conditions for petrification can result. The same phenomenon is caused by embedding the wood in volcanic ash and tuffite following a volcanic eruption. Without contact to the oxygen in the air, the wood constituents are leached out and replaced by minerals from the surrounding soil.

American scientists have been successful in petrifying wood within a few days (5). During this process, the organic wooden material is replaced little by little by minerals – for example, crystallized silica – so that the original structure is preserved completely (6).

## References

1. Wissenschaftsmagazin MaxPlanckForschung, Edition 2/2006.
2. Larry Vardiman, Andrew A. Snelling, Eugene F. Chaffin. Radioisotope und das Alter der Erde, Hänssler-Verlag, Holzgerlingen (2004):189–227.
3. Thomas Herzog. Schnelle Erdölbildung durch hydrothermale Prozesse – Naturnahe Modellierung der Hydro-Pyrolyse und Beispiele aus der Lagerstättenkunde, Studium Integrale (April 2003): 20–27. http://www.wort-und-wissen.de/index2.php?artikel=sij/sij101/sij101-3.html.
4. Borys M. Didyk und Bernd R.T. Simoneit. "Hydrothermal oil of Guaymas Basin and implications for petroleum formation mechanisms." *Nature* 342 (2 November 1989): 65–69.
5. Yongsoon Shin, et al. "Pacific Northwest National Labours, Richland." *Advanced Materials* 17, 73.
6. Hamilton Hicks. Mineralized sodium silicate solutions for artificial petrification of wood, US Patent Number 4,612,050 (16 Sept. 1986): 1–3.

# COSMOLOGY AND THE BIG BANG THEORY

In research matter and the cosmos are closely linked together. In the standard model of conventional physics it is assumed that the universe in which we live and the matter of which we are made originated from a big bang. However we are a long way from understanding what matter actually is, and the big bang theory has to get along with only four percent visible matter, while the rest probably consists of dark matter, dark energy and dark flow.

## An open letter to the scientific community

The big bang theory relies on a growing number of hypothetical entities, things that we have never observed. Many claims in the standard model of the big bang theory are contradictory. For this reason, more than 500 natural scientists have addressed an open letter to the scientific community expressing criticism on the open questions surrounding the big bang theory. These included world-known scientists such as Halton Arp, Hermann Bondi, Thomas Gold, Jayant Narlikar and many others (1).

Yet, the big bang is not the only framework available for understanding the history of the universe. Plasma cosmology and the steady state model both hypothesize an evolving universe without beginning or end. These and other alternative approaches can also explain the basic phenomena of the cosmos. Supporters of the big bang theory may retort that these theories do not explain every cosmological observation (which is also the case with the big bang theory). That is scarcely surprising, as their development has been severely hampered by lack of funding. It is therefore difficult to compare these alternatives with the well-studied standard model of big bang.

## Reference

(1) *New Scientist*. (22 May 2004). http://www.cosmologystatement.org.

# 52

# SINGULARITY AND INFLATION

In the context of cosmology, the existence of the universe started with a big bang. All matter and energy, all space and time were supposedly ignited in a single spot with infinite temperature and density, the so-called singularity. However, today, there is no known mechanism which could lead out of such a singularity. Moreover, it remains unclear whether the natural laws known today came into existence before, during or after the inflation following the singularity. Generally, the border character of scientific questions concerning origin implicates many uncertainties.

The big bang specialist Joseph Silk believes that "based on rational assumptions, a singularity in the past is unavoidable" (1). With this he means a concentration of matter, energy, space and time, i.e., something which escapes any scientific test ore mathematical modelling.

In the history of philosophy, a simple question concerning God was treated for many centuries: "If God created everything, who created God?" This question would have to be asked in the same manner concerning the singularity. It is obvious that it is not possible to answer either question on a scientific basis.

The famous physicists Stephen Hawking wrote, "In the case of the singularity (at the beginning), general relativity and the other physical laws were not effective. It is not possible to predict what will emerge from this singularity" (2).

## Inflation

In addition to the uncertainty of what will evolve from a singularity, the question also remains regarding a mechanism which could have led out of this singularity (3). During inflation that follows the singularity according to the standard model, the universe supposedly expanded at a rate greater than the speed of light during the first fraction of a second. It is not possible to reproduce this process with the natural laws known today (4). It is also unknown how this inflation could have stopped.

## Law of conservation of energy (first law of thermodynamics)

One of the best-proven natural laws is the law of conservation of energy. It states that matter and energy are never lost nor can they be created. Matter, heat, electricity, light, and sound are various forms of energy that can be converted from one form into another, but never get lost or created anew.

The natural laws known today make no exception. Regardless of whether the universe was created by God, or by a singularity and inflation, the formation of matter/energy in the universe contradicts the natural laws known today.

## References

1. Joseph Silk. *The Big Bang*. (New York: W.H. Freeman & Co, 2001): 397.
2. Stephen Hawking. *A Brief History of Time*. London: Bantam Press, 1988): 122.
3. Alex Williams, John Hartnett. Dismantling the Big Bang, Master Books (2006): 13.
4. Ref 3., 117.

# 53

# FORMATION OF GALAXIES

Following the inflation, minor irregularities in the gas density supposedly led to conglomerations leading to the formation of galaxies. However, there are many open questions concerning the formation of galaxies. The big bang theory is still considered a hypothesis.

Approximately one second after the big bang, stable atom nuclei would have been formed. During the next 100,000 years, the universe continued to expand while the temperature decreased and the electrons united with the protons to develop normal atomic structures.

Well-known big bang specialist Joseph Silk wrote (1):

"The big bang theory has not solved three basic problems to date:

1) What happened before the beginning?
2) The nature of the singularity itself
3) The origin of the galaxies."

In the past decades, a number of theories have been advanced, attempting to clarify the origin of galaxies based on the big bang theory. However, none of these was capable of convincing the experts. Within the scope of the big bang theory, the formation of galaxies cannot be explained (2).

# References

1. Joseph Silk. *The Big Bang*. (New York:W.H. Freeman & Co, 2001): 3. Auflage, P. 385.
2. Alex Williams, John Hartnett. *Dismantling the Big Bang*. (Green Forest: Master Books, 2006): 128.

## 54

## FORMATION OF STARS

The origin of the stars is the heart of cosmology. Stars are the source of solar energy and, according to the big bang theory, the only sources in which the heavy elements in the universe (metals) could have been formed. However, in spite of continuing affirmations from many cosmologists, the origin of the stars is still unsolved.

Stars are glowing balls consisting primarily of hydrogen held together by their own gravitational force. Allegedly, they developed because of minor irregularities in the expanding hydrogen following the big bang. The problem is that any accumulation of gases will increase their temperature. Heat causes increased internal pressure, and this will stop further accumulation.

After the accumulation of the gas has come to a standstill, it would have to cool down. Accumulation could continue only after the hydrogen would have cooled down. However, one single cooling phase would require up to forty billion years, while the universe is allegedly only fifteen to twenty billion years old.

### Exceptions

In gas clouds, up to approximately ten times heavier than the sun, the development could progress at a much higher rate. Since its gravitational force would be considerably higher, the high temperatures would develop much more quickly. After only one million years, they would have used up the hydrogen and become "red giants." After all further possible nuclear reactions have taken place, a gigantic explosion would occur resulting in a supernova. The external part of the star would have been blown out into space and the inner part would become a neutron star (1).

If the gas cloud were heavier than ten times the sun, the red giant phase could have been reached after approximately one million years, resulting in an even greater catastrophe: When the core collapses, the field of gravity would become so large that even the neutrons in the individual atoms would collapse. It is assumed that the star would then become a so-called black hole 2.

# References

1. Alex Williams, John Hartnett, 140–142.
2. A.K. Kembhavi und J.V. Narlikar. *Quasars and Active Galactic Nuclei.* (Cambridge, NY: Cambridge University Press, 1999): 101–103.

# 55

# ORIGIN OF PLANETS

An attempt has been made to use computer simulations to explain how gas planets, rocky terrestrials and ice planets could have been formed. It is still an open question how a disc of dust (as allegedly surrounded our sun) could have accumulated to create planets. The known gravity is insufficient to accomplish this. Moreover, the orbits of the planets and moons in our solar system do not have a random structure, but they follow exact mathematical laws.

The process of a supernova explosion, in which a star such as the sun develops and heavy elements such as iron, nickel, and lead are formed, can be simulated. It is also possible to calculate how a gas and dust disc could have been formed. However, it is still unclear and highly controversial whether planets could have formed from such a gas and dust disc and how that could occur (1).

## Gas planets

Computer simulations of the formation of our solar system showed that gas planets do not form in a disc surrounding a star, because the gravity is far below the limit required for accumulation of the gas. Jupiter's mass is approximately 1,000 times less than the sun. If it is not possible even for the sun to accumulate by gravity, it is even less imaginable that the mass of Jupiter would suffice to accumulate. According to theory, a number of supernova explosions would have been required to achieve it.

## Rocky terrestrials

To explain the formation of rocky terrestrials, it has been proposed that a number of meteorites could have been aggregated. However, meteorites do not consist of dust, but rather solid rock or iron. Moreover, the meteorites themselves have too little gravity to accumulate.

## Ice planets

The development of ice planets is even more difficult to explain. They would have to manage with very little material so that accumulation would require an extremely long time.

## Precious metals on Earth

According to the common theory on formation of the Earth's crust, there should be no precious metals on the surface of the Earth. Precious metals such as gold, platinum, and iridium join under certain conditions readily with iron. Therefore, they would have migrated slowly during the molten state into the iron-rich core during the hot primordial period which existed allegedly for a million years.

In order to support the conventional model of development of our planet, it has been proposed that all deposits of precious metals close to the surface originated from impacts of metallic meteorites (2). The conventional model of development of the planetary system is rarely questioned.

## References

1. Alex Williams, John Hartnett, 151-155.
2. Gerhard Schmidt, at European Planetary Science Congress in Münster, 22 Sept. 2008.

# 56

# SURFACES OF PLANETS AND MOONS

If the planets and moons in our solar system have developed from a more or less homogeneous gas or dust disc, the following question comes up: Why are their surfaces composed of radical different materials? Why are there no identical planets or moons? This questions the current theory of the development of planets and moons.

The enormous variety of surfaces of planets and moons in our solar system shows impressively that the scenario according to which these heavenly bodies purportedly originated from a homogenous gas or dust cloud is not realistic (1). The better the data become, the clearer the differences appear (2).

## A few examples

The surfaces of Jupiter and Saturn consist primarily of liquid hydrogen and helium; however, each has a different composition. The surface of Venus is covered by a dense atmosphere of carbon dioxide and sulfuric acid. The surface of Mars is similar to a dry rocky desert on the Earth.

The surface of the moon Europa is remarkably uniform and shows hardly any meteorite craters at all. New data indicates that this moon has a surface which is full of aggressive, corrosive substances. The surface of the Earth's moon is a dust desert. The Jupiter moon Io has a surface consisting of sulfur and sulfur hydroxide. The Saturn moons Enceladus and Tethys are covered with aqueous ice. Saturn's moon Titan is covered with liquid ethane and methane.

## Conclusion

The celestial bodies in our solar system appear to be well formed and are individually distinguished. The question of whether they could have formed from a homogeneous cloud of gas or dust cannot be answered.

It is possible that the planets and moons to be researched during the coming decades will emphasize the impressive differences of the celestial bodies. Our solar system, the Milky Way and the rest of the universe are possibly composed as functionally as, for example, the human body. Could it be that each one of the heavenly bodies fulfils a specific purpose?

## References

1. Kendrick Frazier. *Das Sonnensystem.* (Time-Life Books, 1985): 128–145.
2. Norbert Pailer und Alfred Krabbe. Der vermessene Kosmos, Hänssler (2006) : 99–116.

## 57

## PRECISION PLANETARY SYSTEM

The planetary system surrounding our sun is structured in a very precise manner. However, even minor changes which inevitably occur can lead to situations in which some planets could come into a chaotic orbit after only ten million years. This means that eventually the planets would be lost in the expanse of the universe or would crash into the sun. It is by no means certified that the solar system exists since 4.5 billion years.

Since Newton discovered the laws of gravity in 1683, we know that the planets orbit around the sun in a very stable manner. They behave as regularly as a clock. Moreover, the distance to the sun follows a mathematical rule (1). It is also remarkable that Venus has a retrograde or opposing self-rotation, contradicting the dust disc theory.

It is proposed that the retrograde self-rotation happened after a collision with an asteroid or meteor, but this is highly unlikely because it would have influenced the path of Venus around the sun. The fact that the path of Venus is very close to a circle (the smallest eccentricity of all planets) argues strongly against.

### Long-term stability

Doubts on the long-term stability of our solar system arose as computer specialist Gerald Jay Sussman and astrophysicist Jack Wisdom proved. They used a special computer built just for such calculations. They found that the planet Pluto had already assumed a chaotic orbit because of interference from other plants in our solar system.

Sussman and Wisdom simulated the motion of all the planets and ascertained that the smaller planets should assume chaotic orbits after approximately fifty million years (2). These calculations were confirmed by the French theoretician Jacques Laskar who obtained nearly the same results in 1990, independent of Sussman and Wisdom (3) (4).

Therefore, the assumption that the planets in our solar system have been orbiting stably for 4.5 billion years, the alleged age of the solar system, has to be considered critically (5).

## References

1. Henry M. Morris. *Men of Science, Men of God.* (San Diego, CA: Creation-Life Publishers, 1982): 44–46.
2. Gerald Jay Sussman und Jack Wisdom. "Chaotic Evolution of the Solar System." *Science* 257 (3 July 1992): 56–62.
3. Jacques Laskar. "A numerical experiment on the chaotic behaviour of the Solar System." *Nature* 338 (16 March 1989): 237–238.
4. Jacques Laskar, 266–291.
5. Hansruedi Stutz. "Chaos im Sonnensystem." *factum* (January 1993): 43.

# 58

# EARTH TO MOON DISTANCE

The moon orbits the Earth and its gravitational attraction is responsible for the tides in the world's oceans. Gigantic masses of water are pushed back and forth requiring a great deal of energy. The moon supplies this energy by moving away from the Earth by 3.8 cm each year. Even if the Earth and moon had originally touched one another, this process could therefore have continued for a maximum of 1.3 billion years. This is too short for the alleged 4.6 billion year old Earth/moon system.

Over a century ago, the astronomer George Darwin, son of Charles Darwin, discovered that the moon was moving away from the Earth in a slow spiral. The reason for this is the mutual tide effect of Earth and moon. The moon moves away from the Earth at a rate of 3.8 cm per year.

Although this value is small, it cannot be ignored over long periods. Interesting in this respect is that the tide effect shows a very pronounced function of the distance between Earth and moon. For this reason, the variations must have been much greater in the past as the moon was closer to the Earth, than it is today (1).

Calculating the time for the moon to get to the present distance, we get 1.3 billion years (2). It would have been in contact with the Earth 1.3 billion years ago. Moreover, one billion years ago, it still would have been so close to the Earth that it would have caused extremely high tides. The geologic formations should have to show if there had been such high tides, but they show nothing.

## Stabilization of Earth's axis

The moon causes the tides on the world's oceans; however, it also contributes to the stabilization of the Earth's axis. Jacques Laskar found that the Earth's axis could fluctuate by up to eighty degrees if the moon with its relatively high mass would not stabilize it. The angular position of the Earth's axis appears to be quite stable at 23.3 degrees.

## References

1. Danny R. Faulkner. "The current state of creation Astronomy." Proceedings of the Fourth International Conference on Creationism (1998): 208.
2. Don B. DeYoung. "The Earth-Moon System." Proceedings of the second International Conference on Creationism, Pittsburgh, USA (1990): 81.

# 59

# PLANETARY RINGS

The planetary rings of all four gas planets are demonstrably short-term phenomena. They cannot be older than 10,000 years. Since they must not have existed in context with the planets from the beginning, it is possible that the planets themselves are older. However, it is notable that the planetary rings are observed simultaneously on all four gas planets in the solar system. The uncommonly sharp delimitation of the rings is also astonishing. Since the particles in the ring continuously collide with one another, the edges should be smeared in the course of time.

One of the four gas planets is Saturn. It is surrounded by several thousand rings, which can be categorized into seven primary rings. The entire extent of the ring system is larger than the distance between Earth and moon. The manner in which these rings were formed by natural processes is a complete mystery (1). However, they cannot be very old because they disintegrate during only a few millenniums due to continuous loss of material (2).

## Different properties

Saturn's rings consist of objects, which are up to several meters in size. However, there are significant differences between the individual rings. Only few particles smaller than 5 cm are present in the B ring and the inner areas of the A ring, while they occur more frequently in the C ring and outer A ring.

Further rings have been found in the inner and outer areas of the B ring, which are several hundred kilometres wide and contain highly differing quantities of material. A thick, 5,000 km-wide core contains a number of bands in which the density is four times higher than in the A ring and nearly twenty times higher than in the C ring.

The chemical analysis of the A ring indicates unexpectedly pure grains of ice containing certain silicate admixtures toward the centre. In contrast to the relatively unstructured A ring and the wavy structure present in the C ring, the B ring shows numerous additional structural characteristics (3).

## Conclusion

The notion that these rings developed through natural processes is hardly plausible. They give the impression that for some reason they were formed and structured in precisely the manner we see them today.

## References

1. Stephen Battersby. "First images of Saturn's rings bring surprises." *New Scientist* Nr. 2455 (10 July 2004).
2. Norbert Pailer und Alfred Krabbe. Der vermessene Kosmos, Hänssler (2006) : 136.
3. Hans Zekl. Cassini: Der Stoff, aus dem die Saturnringe sind, astronews.com (30 May 2005) http://www.astronews.com/news/artikel/2005/05/0505-020.shtml.

# 60

# SHORT-PERIOD COMETS

Our solar system contains far fewer short-period comets with an orbital period of between twenty and two hundred years (Halley-type comets) than comets with an orbital period of less than twenty years (Jupiter comets). Only one percent of the Halley-type comets anticipated according to calculations can actually be observed. This is far too low for the concept of a billion-year-old solar system into which new comets enter continuously. They then orbit around the sun, starting from long-periods, in continuously shorter orbits.

Comets are small and irregularly shaped celestial bodies consisting of gaseous and solid particles. The actual body, the so-called nucleus of the comet, consists of ice (frozen carbon monoxide, carbon dioxide, methane, and aqueous ice) and dust and is frequently compared with a dirty snowball. Comets can have diameters from approximately 1 to 100 km. They move in highly eccentric orbits which can bring them very close to the sun and then sling back out far into space.

As a comet approaches the sun, it heats up and the ice on its surface evaporates. This results in development of the sensational tail. During each orbit around the sun, the comet loses the material of its tail. At the end, it disintegrates completely.

The short-period comets have a life expectancy of 50,000 to 500,000 years. This poses the question of why we still have in our planetary system (which is allegedly billions of years old) so many short- period comets. To solve the problem, a theoretically present Oort cloud is proposed which allegedly continuously supplies new comets.

## Oort cloud

In 1950, the astronomer Jan Hendrik Oort postulated that our solar system could be surrounded by a cloud containing many billions of small comets. He picked this proposition up from astronomer Ernst Öpik's theory advanced in 1932. Oort believed that a star occasionally passes by our solar system throwing one of the comets out of its orbit and catapulting it into the interior of the solar system. However, there is

- a) no direct verification for the Oort cloud. Its existence is purely theoretical, and
- b) if such a cloud did exist and a comet was now and then actually deflected into the inside of our solar system, it is nevertheless highly improbable that it would enter into a short-period orbit.

## Fading problem (1)

The gravitation of the large planets influences the orbits of new long-period comets with an orbital period of greater than 200 years which enter the solar system, to change to such an extent, that they either would be hurled out of the solar system after the first pass or enter into a significantly closer orbit.

If new long-period comets had appeared continuously for hundreds of thousands of years, we would expect to find a large number of comets of the Halley type. Nevertheless, very few Halley type comets have been observed.

## Prograde and retrograde comets (2)

It is not possible to explain satisfactorily the large difference in the frequency of occurrence of the various types of comets. For example, the frequency of prograde comets (those rotating clockwise) and retrograde comets (those moving in the opposite direction) does not agree with the calculations. The ratio among long-period comets observed is around at 50:50.

Dynamic calculations indicate that prograde comets have a significantly higher chance of being catapulted out of the solar system by the large planets. Therefore, we expect to find approximately twice as many retrograde comets as prograde. However, the ratio is 50:50. This indicates that the prograde comets have been exposed to the danger of being thrown out of the solar system for only a few thousand years.

## References

1. Peter Korevaar. Die rätselhafte Oortsche Wolke, Studium Integrale 2002/9, 79–82.
2. Paul A. Wiegert. "The evolution of long-period comets." Dissertation, University of Toronto, 1996.

# 61

# SUPERNOVA REMNANTS

A supernova remnant (SNR) is an expanding cloud of dust and gas. An SNR should be perceivable for more than a million years before it disintegrates. However, the number of supernovae in our Milky Way is considerably lower than expected. The number of SNR agrees with the Milky Way, which is approximately 7,000 years old.

When a star with approximately twenty-five times the mass of our sun has burned a sufficient quantity of hydrogen to helium, it explodes. This temporary gigantic release of energy leads to an extraordinarily bright light within the course of a few days or weeks that can outshine all the other stars in the same galaxy. Such an event is called supernova.

A supernova can release the amount of energy that would normally be radiated by 1,000 suns over a period of eight million years (1). It leaves a gigantic cloud of gas, the supernova remnant (SNR), and a small central star. The SNR continues to expand following the explosion at a rate of over 7,000 km/sec and can reach a diameter of several light years in the course of time.

## The SNR expansion process is described in three stages

1. During the first 300 years, the SNR expands to a diameter of approximately twenty-three light years.

The gaseous SNR then changes slowly into a liquid state.
2. During the next 120,000 years, the SNR should continue to expand to a diameter of approximately 350 light years. During this process, the now liquid droplets slowly form to a solid dust.
3. During the next six million years, the SNR would then thin out because of the expansion so that it finally could no longer be distinguished from its surrounding.

Approximately every twenty-five years we can observe a supernova in our Milky Way. Depending on its position in the galaxy, the light from the SNR is weakened more or less by interstellar dust so that some of them are not visible any more.

## Calculations and observations (1)

First stage SNR According to calculations, 19% of twelve first stage SNRs should be visible; two have been observed.

Second stage SNR According to calculations, 47% of 4,800 second stage SNRs should be visible. However, only 200 have been observed. This number we expect after approximately 7,000 years.

Third stage SNR According to calculations, 14% of 40,000 third stage SNRs should be visible. However, none whatsoever has been observed. This agrees with an age of the Milky Way of about 7,000 years.

## Crab nebula

When the first photographs of the crab nebula were taken in the beginning of the twentieth century, it was noted that the nebula was expanding. Calculating back using this expansion rate, it was possible to conclude a supernova explosion 900 years ago. In fact, a supernova did occur in 1054, which was observed and is documented in thirteen independent historical sources (2).

## Swan nebula (Cygnus)

Calculations performed some time ago indicated that the swan nebula is 100,000 years old. However, new data indicates that this figure should be reduced to fewer than 3,000 years. One of the variables on which the expansion rate of a nebula depends is the density of the interstellar medium. Near the swan nebula, this density is approximately ten times lower than the standard density in space. For this reason, new calculations revealed that the swan nebula has expanded to the size observed today in fewer than 3,000 years (3).

## References

1. Keith Davies. "Distribution of Supernova Remnants in the Galaxy." Proceedings of the Third International Conference on Creationism, Pittsburgh, Penn., USA (1994): 177.
2. Jonathan Sarfati. "Exploding stars point to a young universe." *Creation ex nihilo* 19/3 (June-August 1997): 46–48.
3. Keith Davies. "The Cygnus Loop – a case study." *Journal of Creation* 203 (2006): 92–94.

# 62

# METALLICITY OF DISTANT OBJECTS

According to the big bang theory, all objects in the universe originally consisted of hydrogen and helium. Heavier elements formed slowly over billions of years because of supernova explosions. Nevertheless, there is no systematic difference in the metallicity (frequency of metals) between distant objects and close objects. This contradicts the big bang model. The light, which we see today from distant celestial bodies, should have been travelling for billions of years according to the big bang theory before it reached us, offering a glimpse into the remote past.

Frequently it is said that the light reaching us from distant objects allows a glimpse into the past of the universe. This light has allegedly been travelling for many billions of years before it reaches us. The systematic differences in the metallicity we expect between remote and near objects according to the big bang theory are, however, not clearly detectable. (1) (2). If the light (e.g., the spectrum) from celestial bodies is analyzed, it is possible to estimate quite precisely the quantity and quality of elements present in the specific celestial body. The term metallicity is a common designation for the presence of elements heavier than hydrogen and helium. According to the big bang theory, hydrogen and helium developed to continuously heavier isotopes in the stars in a process lasting billions of years. Assuming that no metals were present at the beginning, conclusions can be drawn regarding the age of an object.

The distance of the galaxies is determined based on the red shift in the light from these galaxies. According to the big bang theory, it should be possible to identify young galaxies during their initial phase of development. New measurements have, however, shown that there is no significant difference betweens the metallicity of galaxies close to Earth (i.e., old galaxies) and remote galaxies (i.e., young galaxies) (3).

## Rotating universe

According to the theory of relativity, light is subjected to a red shift when an object moves crosswise (transversely) to the observer (4). It is therefore possible that the universe is considerably smaller than assumed according to the big bang theory and that it rotates around an axis that possibly extends through our Milky Way. If this were true, the estimated age of the universe would have to be reduced drastically.

## References

1. Anna Frebel. Auf der Spur der Sterngreise, Spektrum der Wissenschaft (Sept. 2008): 24-32.
2. Peter Bond. "First stars seen in distant galaxies." Royal Astronomical Society Meeting, 5 April 2005.
3. Norbert Pailer und Alfred Krabbe. Der vermessene Kosmos, Hänssler (2006) : 64-66.
4. Andreas Müller. wissenschaft-online.de, August 2007, http://www.wissenschaft-online.de/astrowissen/lexdt_d02.html.

# 63

# ANTHROPIC PRINCIPLE

The anthropic principle refers to the incredible fine-tuning of the various natural constants. If even one of the over forty known natural constants was to deviate minimally from its present value, life would not be possible on the Earth. Some scientists who do not believe in an intelligent creator of the universe solve this dilemma with the so-called multiversum theory that postulates an infinite number of universes. Further, it also claims that we live in precisely that universe in which life is possible. Naturally, it is possible to prove everything and nothing with such a theory.

The probability that our universe corresponds to precisely one specific manifestation is approx. $1:10^{62}$ in the opinion of some physicists. This corresponds to:

0.000'000'000'000'000'000'000'000'000'000'000'000'000'0 00'000'000'000'000'000'000'001% (1).

One physicist clarifies the precision of the fine tuning of natural constants in the following manner. Imagine a two-Euro coin which had to be hit with a rifle bullet. The rifle would have to be positioned at one end of the universe and the coin at the other end!

## A few examples

If the gravitational constant would be slightly lower, this would prevent stars such as the sun from starting nuclear fusion. If it were only slightly higher, the energy stock of the stars would be used up within a very short time.

If the nuclear forces holding atoms together were slightly higher, the electrons would crash into the nucleus as if it were a black hole. If they were only slightly weaker, chemical reactions would not take place. For example, water would not have such particularly conspicuous anomalies (freezing, boiling point, density curve, etc.) and life on a water base would be impossible (2).

Astronomer Martin Rees has selected and described six of the many natural constants indicating how none of them can deviate even slightly from the existing value. Otherwise, life would not be possible on Earth (3).

## Multiuniverse or eternal lord of creation?

Rees assumes that there must be an infinite number of universes and that one of these has precisely the right natural constants by chance. It is therefore possible to believe in a theory propounding an infinite number of universes or to believe in one single infinitely intelligent and omnipotent creator of the one universe in which we live (4).

## References

1. Peter C. Hägele. Das kosmologische anthropische Prinzip, Kolloquium für Physiklehrer, Universität Ulm, 11 Nov. 2003, http://www.uni-ulm.de/~phaegele/Feinabstimmung_Physik.pdf.
2. Hansruedi Stutz Das anthropozentrische Prinzip: Der Mensch im Mittelpunkt des Universums, factum (July/August 1991): 39.
3. Martin Rees. *Just Six Numbers*. HarperCollins Publishers, 1999.
4. David Tyler. "Parallel Universes: Has God anything to say?" *Origins* 34 (March 2003):14–15.

# 64

# MICROWAVE BACKGROUND RADIATION

The cosmic microwave background radiation coming to us from all directions in space is much more uniform than would be expected based on the big bang theory. With the introduction of dark matter, it was finally possible to explain the small measured irregularities with the predicted values. However, the existence of dark matter is still speculation.

In 1926, Sir Arthur Eddington argued that all celestial bodies are bathed in star light and that interstellar space must therefore have a temperature of approximately 3 K (-270°C) (1). After him, George Gamow interpreted this cosmic background radiation as an afterglow from the big bang. He calculated a figure of 5 K (-268°C). In 1961, he revised his figures predicting 50 K (2). In 1964, the two astronomers Arno Penzias and Robert Wilson finally measured a temperature of 2.7°K.

Later, NASA constructed a special satellite to map the entire cosmos using microwave background radiation. According to the big bang theory, the irregularities in the expansion of hydrogen and helium should have caused the large structures in the universe. However, the first satellite's instruments proved to be too insensitive to measure any differences at all.

Then, a new satellite was built equipped with instruments, which were thirty times more sensitive. Subtle differences were, in fact, then noted. However, unpleasant surprises were also found. For example, it was found that the cosmos has a North and South Pole and an Equator (3). This, for its part, could mean that the universe has a centre and that we are close to this centre. In the big bang model, these results could not be understood because it is assumed that the universe does not have a centre.

Moreover, astronomer Tom Van Flandern believes that the absorption of microwaves in intergalactic medium and the absence of the effects of gravitational lenses contradict the big bang model (4).

Considering the erroneous predictions of the big bang theory (the exceptional uniform structure of microwave background radiation) and the alternative interpretations of the microwave background, it is necessary to critically consider the big bang theory in its entirety.

## References

1. Arthur S. Eddington. *The Internal Constitution of the Stars*. (New York: Dover Publications, 1926, republished 1959): 371.
2. Tom Van Flandern. "The Top Thirty Problems with the Big Bang." *Apeiron* 92 (2002): 72–90.
3. David Whitehouse. "Map Reveals Strange Cosmos." BBC News (3 März 2003). http://news.bbc.co.uk/go/pr/fr/-/1/hi/sci/tech/2814947.stm.
4. Tom Van Flandern. *Dark Matter, Missing Planets and New Comets: Paradoxes Resolved, Origins Illuminated*. (Berkeley, CA: North Atlantic Books, 1993): 100–107.

# PHILOSOPHY

In modern science, naturalism has become the leading paradigm. A doctrine can be called naturalistic if nature alone is the basis and standard for all phenomena. The naturalistic approach resulted primarily from the motivation to delimit from supernatural phenomena in the religious sense. The existence of miracles, supernatural beings, and spiritual insights is rejected.

Evolution, primordial soup, and the big bang theory are of great significance for the naturalistic worldview. However, the naturalistic ideology must be questioned on the ground of some philosophical considerations. For example, in evolution theory, the term "chance" in the sense of aimlessness, purposelessness, and pointlessness is an imprecise allegation without any meaning.

Why are signs of teleology (purposefulness) and planning to be found everywhere in the universe? How can humans question the purpose of life? How can beauty with no apparent purpose and natural perfection be explained? These and other questions remain unanswered in the dogma of evolution theory.

# 65

# PARADIGM OF EVOLUTION

Polls have shown that many people are of the opinion that evolution theory is a scientifically- proven fact. Few know that scientists operate with provisional models (verification) and contradictions (falsification). The prevailing patterns of thought (paradigm) of evolution, primordial soup and the big bang theory are of philosophical origin (enlightenment, rationalism/ naturalism) and cannot be proven with the methods of natural science.

In popular scientific publications, it is frequently maintained that the theory of evolution is a proven fact. This claim is not tenable on the basis of natural science. If we eliminate supernatural intervention by an intelligent creator from the very beginning and accept the models of evolution, primordial soup and the big bang theory as our paradigm, it is not possible to conclude that no creator exists, because we have excluded a creator from the very beginning.

## Empirical and historical science

Scientific theory distinguishes between empirical (experimental) and historical science. In both sectors, an attempt is made to first setup general explanations (hypotheses). Then, it is determined whether the prognosis derived from it actually applies.

In empirical science, verification can be accomplished by experimentation and observation. Perception and scientific theoretician Karl R. Popper explains: "The activity of the scientific researcher is to establish theorems or systems of theorems and to verify them systematically; in the empirical sciences, hypothesis and theoretical systems are established particularly and checked on the basis of experience made with observation and experiment" (1).

In the historical sciences (which includes research on evolution, primordial soup and the big bang theory), this is not possible. The various interpretations have to be verified primarily according to plausibility criteria.

## Conclusion

The models of evolution, primordial soup and the big bang theory have not progressed beyond the status of a hypothesis. Even if we were successful in producing life in the laboratory, this does not mean that it was possible in the past without intelligence and state-of-the-art human technology.

## Reference

1. Karl R. Popper. Logik der Forschung, Wien, 1934, 1. Kapitel, http://www.ploecher.de/2006/11-PA-G1-06/Popper-Logik-der-Forschung-kurz.pdf.

# 66

# NATURALISTIC WORLD VIEW

With the aid of evolution, primordial soup and the big bang theory, an attempt is made to explain the world in a purely natural manner. However, nature and naturally are very flexible words. At a closer look, it is not possible to distinguish between "natural" and "supernatural."

Many scientists believe that the term "natural" initially has only one single meaning, which differs when used in practice. Nature is first limited to a universe consisting of particles and forces. This excludes gods, angels and all other superstitious objects. Then, however, they turn around and use concepts for rationality and morals that cannot be reduced to particles and forces.

Let us consider the following non-physical variables used in scientific literature: forces, acting from a distance, singularity, infinite, consciousness, intellect, extraterrestrial intelligence, placebo effect, unobservable phenomena such as the interior of stars, dark material, dark energy, quarks, superstrings, the big bang theory and the origin of life. Some scientists even postulate parallel universes or an infinite multiuniverse. How natural is all this? (1)

Then we work with numerous concepts such as the following: information, mathematics, laws of logic, philosophy, history, reason, the scientific method, rationality, classification, causality, induction, and objectivity. Science itself is a concept.

Other non-physical concepts include the moral categories of truth, honesty, ethics, integrity and justice which are timeless and universal and refer to absolute values. Science is dependent on a number of things that cannot be explained with particles and forces.

The scientific journal *Nature* wrote that science adheres to truth even when it is uncomfortable or painful. "The belief of most people tends to reinforce their own interests. This morbid fact is the great strength of science" (2).

Mathematician and logician Kurt Gödel was capable of proving that mathematics cannot verify itself. The physicist David Wolpert recently extended this argument to cover all scientific argumentation. According to the words of physicist Philippe M. Binder, Wolpert was successful in proving that "the entire physical universe cannot be understood completely by one single system of conclusions existing within it" (3). Therefore, the naturalistic worldview is not capable of substantiating itself from itself and within itself contrary to the desire of many of its advocates.

# References

1. David F. Coppedge. *Acts and Facts* 38/4 (April 2009): 19.
2. "Nature, Editorial, Humanity and Evolution: Charles Darwin's thinking about the natural world was profoundly influenced by his revulsion for slavery." *Nature* 457 (12 February 2009) 763–764.
3. Philippe M. Binder. "Philosophy of science: Theories of almost everything." *Nature* 455 (16 October 2008): 884–885.

# 67

# DOGMA OF EVOLUTION THEORY

In the mind of many scientists, science is nothing more than applied naturalism or in the words of Steven Weinberg, "Science – regardless of which branch – can progress only when it assumes that there is no divine intervention and recognizes how far it can go with this assumption." However, the existence of God cannot be excluded based on natural science. And, if God did exist, science – regardless of which branch – would make real progress only when He is included in its considerations.

As shown by the introductory quotation from Weinberg (1), many scientists work under the non-verifiable basic assumption that there is no divine intervention in nature. However, science should not allow the assumption of dogmatic prerequisites as a matter of principle. True science means finding the truth, regardless of what means are used and regardless of what they contain.

In spite of many applicable experiments, it is necessary to establish that it is not possible to disprove the existence of God. Moreover, it is an irrefutable principle that no knowledge exists without presuppositions and that our cognitive capabilities are subject to limits (see, for example, Gödel's incompleteness theorem and Heisenberg's uncertainty principle).

Karl Popper described this fact, "Certain knowledge is unattainable. Our knowledge consists of critical guessing, a network of hypotheses, a web of speculations … and our guessing is directed by the unscientific, metaphysical belief that there are laws which we can unveil and discover" (2).

## On the history of evolution theory

The idea of evolution is not new. Even centuries before Christ, there were concepts that life had developed and that living creatures originated from one another. For example, Anaximander of Miletus (610–547 BC) supported the belief that creatures similar to fish had emerged from the waters and evolved into animals and humans. The breakthrough for the evolution concept and its acceptance as well as its social implementation was made possible by philosophical views during the eighteenth century.

As rationalism placed human reason as the highest authority the end of the seventeenth century and materialism made material the only real absolute value, it was possible for the naturalistic philosophical current of thought to develop optimally. Naturalism does not recognize any authority outside the visible world. The Britannica dictionary describes naturalism as, "The doctrine that phenomena are derived from natural causes and can be explained by scientific laws: opposed to supernaturalism" (3).

The philosopher Wilfrid Sellars wrote, "When we speak of describing and explaining the world, the natural sciences are the measure of all things" (4). The logical sequence of this worldview is inevitably a type of development doctrine that refutes all supernatural events. However, in the final analysis, scientific knowledge has not lead to exclusion of the supernatural. On the contrary, it is excluded from the very beginning by the philosophy of naturalism.

The Jesuit and palaeontologist Pierre Teilhard de Chardin wrote that evolution is a "generally valid postulate to which all theories, all hypotheses and all systems will have to bow in the future and must satisfy it in order to be perceived as imaginable and true. Evolution is a light which elucidates all facts, a track which all paths of thought must follow" (5).

Evolutionist and Nobel Prize winner Konrad Lorenz believes that "irrational, emotionally obsessed resistances are exclusively responsible for the fact that there are still educated people who do not believe the theory of evolution" (6).

Zoologist D.M.S. Watson, however, wrote that evolution is accepted, "not because anyone has observed it, or because it has been proven to be correct by a contiguous logical chain of proof, but because it is the only alternative; the act of creation by God is simply unimaginable" (7).

## References

1. Steven Weinberg. *Dreams of a final Theory*. Vintage, 1994.
2. Karl R. Popper. Logik der Forschung, zitiert in Volker Kessler, „Ist die Existenz Gottes beweisbar?" 84.
3. Britannica world language dictionary, vol. 1 (1964): 845.
4. Wilfrid Sellars. *Science, Perception and Reality*. London: Routledge and Kegan Paul, 1963): 173.
5. Marie-Joseph Pierre Teilhard de Chardin. *The Phenomenon of Man*. 1959, deutsche Ausgabe: Der Mensch im Kosmos, C.H. Beck, München, 1959.
6. Konrad Lorenz. zitiert aus Hoimar v. Ditfurth, Evolution, Hoffmann und Campe (1975):13.
7. D.M.S. Watson, *Nature* 123 (29 June 1929): 233.

# 68

# EVOLUTIONARY PSYCHOLOGY

For a number of years, public media have offered many popular scientific contributions in which human and animal behaviour are explained in context with the evolution theory. This frequently involves studies regarding human sexual behaviour. However, many conclusions on the evolutionary formation of cognitive mechanisms have proven to be circular arguments. Others are formulated so vaguely and with so little differentiation that they can only be considered stories which sound plausible; however, they cannot be confirmed or contradicted.

What purpose does the female orgasm serve? Is the frequency of orgasms higher for women who have a partner with a high income? How have mental phenomena such as "affection" and "concern for one's own children" developed? Is our brain the product of a long process of adaptation?

Questions such as these are treated based on evolution theory and can be answered plausibly on various occasions. Philosophers and theologians who believe in God proceed in a similar manner when they support their religious opinions using a model of a world created by God based on the Bible or other documents. In both cases, it is hardly possible to speak of proof in its actual sense.

## Definition and history of evolutionary psychology

Evolutionary psychology is a branch of research that attempts to explain the origin of the human psyche using evolution.

Evolutionary psychology is not limited in terms of contents. On the contrary, it is intended to provide a new methodical approach to psychology as a whole. It is applicable for every branch of psychology (1).

In evolutionary psychology, classical psychological data continue to play a major role. However, this data is supplemented, for example, by assumptions on human evolution, "hunter and collector" studies or economic models. Some considerations go back to Charles Darwin; however, an independent and influential approach to evolutionary psychology evolved only in the early 1990s as a result of cooperation between psychologist Leda Cosmides and anthropologist John Tooby (2).

## An example of creation psychology

If the faces of several hundred women are captured tri-dimensionally, and then an average face calculated, a woman appears who is generally described as immaculately beautiful. The tendency to choose an "attractive" partner could be interpreted as a type of "creation psychology," which would create every type of life according to its kind and in which individuals attempt to maintain their own kind on the average.

This would lead us to expect an evolutionary behaviour that showed no tendency whatsoever or any experimental desire to develop in new, extraordinary directions.

## References

1. Aaron Sell, Edward H. Hagen, Leda Cosmides und John Tooby. "Evolutionary Psychology: Applications and Criticisms" in Lynn Nadel's *Encyclopedia of Cognitive Science* (Hoboken: John Wiley & Sons, 2006): 54.
2. Jerome H. Barkow, John Tooby, Leda Cosmides. *The Adapted Mind: Evolutionary Psychology and the Generation of Culture.* (Oxford: Oxford University Press, 1992).

# 69

# CHANCE PROCESSES

In evolution theory, the generally used term "chance" in the sense of "without plan, objective or purpose" has a negative connotation while the use of the term "chance" in the sense of a stochastic process (see definition below) is an imprecise contention without any substance. Theoretical rejection of a guiding force (e.g., a God) or causality of any kind or the flat assertion of stochastic processes is meaningless. This remains so even when a factor of alleged necessity is added to chance.

The chance factor is a recurrent theme in evolution theory. Chance is decisive or at least partially decisive for mutation as well as selection and in the remaining evolutionary factors (recombination, gene loss, gene multiplication, jumping genes, horizontal gene transfer, separation of a population, etc.) (1).

We speak of chance when an event is not the result of a causal inevitability. However, in common speech the term is also used when an event is not foreseeable, predictable, or calculable in practice. Chance should never be confused with unpredictability or incalculability.

## Definition of chance and stochastic according to the Brockhaus Encyclopedia:

Chance: "That which occurs without recognizable reason or intention, that which may occur however does not have to occur (opposite: imperative)."

Stochastic: "Art associated with guessing ... Generic term for probability theory ... Stochastic includes ...all quantifiable aspects of random events" (2).

Ernst Mayr on chance and imperative: "Unfortunately many oversee the fact ... that ... natural selection is a two-stage process. In the second step, selection is, in fact, decisive for adaptation. However, the first step, the origin of the variation, supplies the material for natural selection and stochastic processes (i.e., chance events) dominate here.... Moreover, it should never be forgotten that chance also plays a significant role in the second step of evolution, consisting of survival and reproduction" (3). According to Ernst Mayr "chance" is an "unforeseeable event" (4).

The term chance is used by proponents of evolution, on one hand, in the sense of a stochastic process and, on the other (at least implicitly), in the more common sense of (according to Mayr explicitly) without plan or objective (5).

In this regard, Charles Darwin wrote, "Up to now, I have used the word "chance" when speaking of changes which occur more frequently in organic beings in the state of domestication and more seldom those in a natural state. Naturally, the word chance is not a proper designation; however, it at least implies our lack of knowledge of the causes of specific changes" (6).

## Chance as negative concept

The use of the term chance in evolution theory in the sense of without plan, objective or meaning is a negative claim without any substance. Theoretical rejection of a guiding force or of a god or causality of any kind is meaningless like every negative assertion. The term used in the doctrine of evolution therefore remains unsubstantial even when it is combined with the alleged factor of necessity because unsubstantial times substantial always results in unsubstantial.

## Chance in the sense of a stochastic process

The use of the term chance as an evolutionary factor in the sense of stochastic processes is nothing more than acknowledgment that no or no precise knowledge of how evolution operates is available. If we assume causalities and they were already explained in theory or even researched empirically, it would no longer be necessary to use the term chance or stochastic processes as an expression of lack of knowledge.

In fact, it is not possible to delimit the initial situation before innumerous mutations for alleged upward development of organisms, nor is it possible to draw any statistical conclusions at the genetic level regarding alleged individual macro evolutionary processes and certainly not regarding a number of such contiguous processes. Common assertions about stochastic processes are and remain absolutely unsubstantial without known initial situation or probability calculations.

Moreover, before a stochastic process can occur, it is necessary to develop the individuals; from them a selection is then possible. It is first necessary to invent and produce a die with six dots before it can be cast. It is not possible to designate something as a stochastic process as long as there is no possibility for selection. Here, it is necessary to differentiate precisely between the origin of something new and development of things that already exist. Evolution theory has no explanation for origin of anything new. Blank statements regarding the stochastic process do not provide any substantial (reasonably founded) explanations for development from things already existing.

## Conclusion (7)

The information associated with chance and every statement associated with the chance factor is unsubstantial. The factors chance times law always result in chance: $0 \times 1 = 0$. As soon as a statement contains a factor of chance, the entire allegation becomes unsubstantial, incomprehensible and improvable.

Asserting a theory whose central statement refutes any supernatural force or any other causality and otherwise claims not to know what is happening, does not, in principle, represent any theory at all.

## References

1. Ernst Mayr. Das ist Evolution, 3. A., (München 2003): 177.
2. Der Brockhaus, Naturwissenschaft und Technik, Heidelberg, 2003.
3. Ernst Mayr, 281, 338, 343.
4. Ernst Mayr, 354.
5. Ernst Mayr, 154, 263.
6. Charles Darwin. Die Entstehung der Arten, übersetzt von Carl W. Neumann, Nikol Verlag Hamburg, (2004): 188.
7. Dieter Aebi, Prozessakte Evolution, Evolution contra Kreation aus juristischer Sicht, Dillenburg 2006.

# 70

# CAUSAL EVOLUTIONARY RESEARCH

Causal evolutionary research has an insurmountable problem with its proof: It is necessary for it to prove chance development according to its own theory, based on experience (description of calculable and predictable sequences), the alleged long periods (during which macro evolutionary developments are to have taken place) and unsubstantiated claims (such as that evolution is "directional," however not "goal-oriented"), protect the theory against falsification.

Exactly the same data is available to proponents and opponents of the theory of evolution. The basics for the specific interpretation of the data available are the same for both. They do not extent beyond current observations or current experiments as well as the current proof of experience (causalities, principles). The past, by contrast, can be reconstructed experimentally, but only to a very limited extent.

## Historical and causal evolutionary research

So-called historical evolutionary research in the areas of comparative biology and fossil research is based solely on the data available today about the remains of dead organisms, including fossils.

Causal evolutionary research, with the areas of speciation through so-called evolution factors (mutation and selection) as well as molecular evolutionary research, attempts to prove the current experience or principles of development based on current data.

This requires that historical evolutionary research interprets present facts to the past while in the case of causal evolutionary research, experience gained today is extrapolated into the past!

When experience is transferred into the past, as well as in the case of retro-interpretation of fossils, the specific conviction plays a decisive role! Observation and belief are mixed inevitably. Since individual pieces of data (e.g., the appearance of a single fossil) and individual experience do not provide a correct picture of the past, it is necessary to collect more data and experience to obtain a correct picture of the past.

Since the data are actual as well as innumerable and the experience is actual as well as extremely complex, their combination to form even an approximately consistent, unambiguous, and coherent theory without the guidelines of a predetermined concept is not possible. The conception of the past therefore develops inevitably from the combination of observation and belief or the combination of the methods of induction and deduction (derivation of general phenomena from individual phenomena and vice versa).

## The problem of proof

It is only possible to prove actual and unchanged facts and experience by empiricism. Since a current factual proof for the origin and development of matter and life in the past is totally absent, proof is possible only indirectly through actual experience assuming the same conditions as in the past. Empirical propositions – just as all physical laws – are characterized by calculable, predictable, precisely defined actual sequences.

A causal evolutionary research therefore has an insurmountable problem with proof. It is necessary for them to prove development by chance, i.e., incalculable and unpredictable development, based on calculable and predictable processes, according to their own theory. This is impossible!

The chance factor makes empirical research on macro evolution, i.e., the common origin and advancement of living creatures, impossible at the very beginning!

## Conclusion (1)

Since chance developments cannot be proven with experience sought with causal evolutionary research, only historical evolutionary research such as palaeontology is suitable for proving the alleged macro evolution. Hereby it is supported by geology and, to a limited extent, archaeology, for establishing a chronological relationship to the dating of the rocks containing fossils. There exists a danger of mutual influence of interpretation of the facts. Historical evolutionary research is not an empirical science.

## Reference

1. Dieter Aebi, Prozessakte Evolution, Evolution contra Kreation aus juristischer Sicht, (Dillenburg, 2006): 9.

# 71

# HOMOLOGOUS ORGANS

Similarly-constructed body parts of many living creatures are called homologies. Some examples include the pectoral fins of fish, the front extremities of tetrapod vertebrates as well as the wings of birds and bats. Moreover, all living creatures known to us today are constructed with the same basic building blocks or proteins. The information carrier DNA is also the same for all living creatures. These similarities could indicate a common origin as well as a common creator.

Every creative intelligence has its own specific handwriting. For example, if we consider the pictures and sculptures of Pablo Picasso, we note similarities and a development. However, no one would even think of saying his works have a common descent. Similarities are no proof of common descent. They simply show that the same basic principle was used for different living creatures.

The same applies for DNA strings. The blueprints for similar living creatures are written manifest with the same genetic code, because this code is optimally for all forms of life.

## Problems in interpreting similarities

The interpretation of homologies as an indicator for common descent is concluded only by analogy; however, this is not inevitable. Many similarities can be explained by the function so that a reference to evolution does not provide additional clarification, but rather represents a circular argument (1).

Similarities as indicators for a common descent can be determined clearly only based on empirical data. As a rule, they are recognized as such only under presupposition of evolutionary hypothesis using the principle of economy. Evolution cannot be proven by similarities.

## Contradictory genealogies

Many homologous organs occur in living creatures that can only be very remotely related with one another according to their alleged lineage. For this reason, the majority of coincidental characteristics must be classified as parallel developments in evolutionary theory, which poses significant problems for clarification. There is no generally objective possibility for differentiating between similarity and parallel development. Frequently, characteristics appear to be distributed in the form of building blocks among different species and higher groups (taxa).

Mature organs, organ systems, individual development paths from the egg cell to sexually mature state and genes frequently support contradictory similarity conclusions. This has led to a crisis in the similarity concept because more and more has become unclear, which could serve to support similarities as an indicator for phylogenetic relationships (2).

## References

1. Reinhard Junker. Ähnlichkeiten – Rudimente – Atavismen, Hänssler-Verlag (2002): 18.
2. Junker und Scherer. Evolution, ein kritisches Lehrbuch, Weyel, 2006, 167–190, 301.

## 72

## NATURAL PERFECTION

When we observe nature, we observe that things are perfectly matched and not in some stage of semi-development. Every living creature, no matter how small, fulfils some purpose; every weed is good for something. There are no incomplete ecological systems; the vast majority of living creatures provide a contribution to the common good of the entire ecological system (with the exception of modern man). All this speaks in favour of life on Earth originating from an intelligent creator.

Many humans do not believe in chance. Somehow, we sense that a higher order dominates over the visible world that permeates us and everything on Earth (1). Somehow, we sense that there is a God who differentiates between good and bad, between right and wrong (2).

American physicist Arthur H. Compton (1892–1962) stated, "For me, faith begins with the awareness that a supreme intelligence called the universe into being and created humans. It is not difficult for me to believe this, because it is incontestable that where there is a plan, intelligence is also present. An orderly, developing universe provides evidence for the truth of the most powerful proposition ever made: "In the beginning, God created the Earth" (3).

## Leonardo da Vinci's proportional study according to Vitruv

Leonardo's well-known pin drawing of a mature man as double figure with outstretched limbs drawn into a square and a circle (see title picture) underscores the natural perfection of the human body. Even the fact that many humans can be centred in a square and a circle is highly astonishing. However, Leonardo's picture contains a great deal more.

## Squaring the circle

Only at the end of the twentieth century was mathematician Klaus Schröer successful in proving that this picture also contains the square of the circle. The proportional study according to Vitruv signals a circle with a surface equal to its square with two points – the middle finger on the vertical side of the square through which leads an imaginary arc. Previous art historians and mathematicians simply oversaw the implied circle.

The proportional study according to Vitruv pictures a method for an effective as well as attractive construction for obtaining the approximate square of a circle with a compass and ruler. This method remained unknown for centuries, lost in the grid work of mathematical history, so to speak, and was possible particularly because Leonardo did not openly explain the method on the sheet with the double figure but rather implied it symbolically instead (4).

The natural perfection of the human organism in terms of form and function is overwhelming.

## References

1. *The Bible*. Rom. 1:19–23.
2. *The Bible*. Gen. 3:22.
3. Arthur H. Compton. Speech on 12 April 1936, *Chicago Daily News*.
4. Klaus Schröer und Klaus Irle. Ich aber quadriere den Kreis …, Waxmann, 1998, 105–111.

# 73

# TELEOLOGY AND ORDERLINESS

According to many scientists, the infinite number of cosmic and biological structures that we can observe today allegedly developed by pure chance. This dogma contradicts the goal-directedness (teleology) and orderliness that are recognizable in all of nature. If nature actually had developed by chance processes only, teleology should not be recognizable.

Most evolution advocates attempt to explain the origin of life using matter and natural laws alone. According to this concept, teleology should not exist in nature. In this context, political scientist and biologist Robert Wesson recognized, "The only question where modern authors have a unanimous opinion is that adaptation (through mutation/selection) is not teleological" (1).

This unanimity can be explained from the common, materialistic approach of scientists. But how can the evolutionist Aldous Huxley describe evolution as a "determined chronologically irreversible process" (2) when at the same time purposefulness is specifically refuted.

Nobel Prize winner Jacques Monod had to admit that "the cornerstone of the scientific method is the postulate of the objectivity of nature ... This postulate of objectivity is essential for science...Particularly objectivity obliges us to recognize the teleonomic character of life, to admit that it follows a plan in its structure and performances. The central problem of biology is this contradiction itself" (3).

This contradiction cannot be avoided with a materialistic ideology. The postulate of complete purposelessness can hardly be sustained by consistent thought on the concrete reality of nature (4).

## Vestiges of God in creation (5)

In his book *Spuren Gottes in der Schöpfung,* Reinhard Junker presents a profound description and detailed analysis of the criticism of the principle ideas of the intelligent design movement. He approaches the subject of teleology in biology from the perspectives of scientific theory, science and theology, whereby the incapability of previous evolutional models to explain this phenomenon is elaborated concisely.

## References

1. Robert Wesson. *Beyond Natural Selection.* (Cambridge, Mass.: 1991) German Edition: *Die unberechenbare Ordnung.* (Artemis & Winkler, München, 1993): 31.
2. Johannes Grün. Die Schöpfung, ein göttlicher Plan, 509.
3. Jacques Monod. Le Hasard et la Nécessité (Paris, 1970): 37f. (German edition: Zufall und Notwendigkeit, München, 1971).
4. Phillip E. Johnson. Darwin im Kreuzverhör, CLV Bielefeld, 2003, 145.
5. Reinhard Junker. Spuren Gottes in der Schöpfung? Eine kritische Analyse von Design-Argumenten in der Biologie, Holzgerlingen, 2009.

## 74

## PURPOSE OF LIFE

The question of the purpose of life cannot be answered based on evolution theory. On the contrary: This question cannot even be posed due to naturalistic considerations. The evolutionist Richard Dawkins wrote, "The universe which we see has... no order, no purpose, no good and no bad, only purposeless indifference."

The words of Dawkins, who described the universe as nothing other than purposeless complacency (1), are by no means meant maliciously or as an expression of resentment, but rather the inevitable conclusion when the theory of evolution is sought through consistently to its end. However, if we turn away from chance toward plan and purpose, we recognize that we are a part of a great cosmic plan in which we have a right to ask what is the purpose of our life?

The scientific method of research is neutral regarding the purpose of life. It allows us to avoid subjective opinions and ideological influences to the greatest possible extent. Since many evolutionists deny any purpose behind the origin of life and replace it with a miracle-producing random process, they do, in fact, take a position. For this reason, it is not possible for them to speak of a natural scientific theory, in the true sense.

According to evolution biologist William B. Provine, the modern understanding of evolution allows the conclusion that there is no final purpose of life (2).

Nobel Prize winner Jacques Monod also wrote that humans should awake from their dream and recognize their total abandonment and radical estrangement, in order to know their place as vagabonds at the edge of the universe (3). Otherwise, it is also possible that one day Monod will wake up and recognize that life does, in fact, has a purpose because it originates from a creator who gave it a purpose.

# References

1. Richard Dawkins. "A Scientist's Case against God." *Science* (Aug. 1997): 892.
2. Spektrum der Wissenschaft, Naturwissenschaftler und Religion in Amerika, Larson/Witham (November 1999): 74.
3. J. Monod, Zufall und Notwendigkeit (1977): 151.

# 75

# UNNECESSARY BEAUTY

The unnecessary beauty occurring in nature is an important criterion for intelligent creation. The naturalistic approach fails to explain the development of unnecessary beauty. Natural selection would favour exclusively practical mutations providing a survival advantage in some manner. Unnecessary beauty would not be favoured or selected according to evolution theory.

There are animals and plants which are unspeakably beautiful (according to human standards) and which appear not to have any advantage in comparison to inconspicuous types of the same species with simple appearance. In his work *The Origin of The Species*, Charles Darwin wrote that it would be a hard shock for his theory if many organic structures were "beautiful only to delight the observer" (1). Simply being beautiful would not provide any advantage for evolution. Darwin was aware of this.

Naturally, a great deal of beauty in nature can be explained in connection with the reproduction behavior of living creatures. In many cases, it does, in fact, appear to have no specific purpose.

Structures are present apparently "only for the sake of beauty."

It would not be necessary for many flowers to be anywhere near as beautiful as we know them. In the vast majority of cases, it would be sufficient to simply produce flower petals with the right colour to attract bees and other insects. Bees do not have eyes like those of humans. The beauty of most flowers, especially orchids, does in fact appear to have no specific purpose.

The impressive peacock's tail with its unnecessary splendour or the wonderful pattern and colours of butterfly's wings are probably not practical; they could even be disadvantageous.

Aeolids are tiny snails that can be observed only with a strong magnifying glass. These snails have an elaborate pattern with vivid colours, even though they themselves do not have eyes to perceive an image. Biologist Adolf Portmann wrote in this regard, "With these colourful snails, we see, as in numerous other cases, complex development processes for visual imagery which cannot be associated with any perceptive organ at all and still appear to be formed for 'sight' in terms of colour and shape" (2).

Unnecessary beauty implies an observing, super-ordinate intelligence that not only pays attention to detail and purpose format, but also considers the creature as a whole and places values on harmony and beauty. The same is true for programming the DNA. Inventing intelligent information of the kind written into the DNA is possible only for an intelligent creator.

## References

1. Digitale Bibliothek Band 2: Philosophy: Charles Darwin, „Die Entstehung der Arten", 423.
2. Adolf Portmann. Meerestiere und ihre Geheimnisse, Reinhardt-Verlag, Basel, 1958, 73.

# INFORMATION THEORY

The strongest form of scientific argument is the application of natural laws in such a manner as to exclude a process or occurrence.

## The natural laws of information and their consequences (1)

All living beings contain an absolutely unimaginable amount of information. The conceptual system evolution could only function if information could be produced in matter by chance processes. This information is absolutely necessary, since all blueprints of all individuals and all complex processes in living cells (e.g., protein synthesis) are controlled by information.

The following eight theses present arguments based on the natural laws of information. These natural laws are derived from observation. They exclude the possibility that any information – including biological information – could possibly have originated from matter and energy without the agency of an intelligent originator. They demand a conscious creator with a will and the ability to bring forth the biological information.

## What is a natural law?

The term natural law is used when the general validity of stated principles concerning the observable world can repeatedly and reproducibly be confirmed. In science, natural laws are the principles that enjoy the highest level of trust.

Natural laws ...

- ...have no exceptions.
- ...do not change over time.
- ...exist independently of their discovery and formulation by humans.
- ...can always be successfully applied to as yet unknown cases.
- ...provide an answer to the question of whether or not a conceived process is possible at all.

This is a particularly important application of natural laws.

The laws normally categorized as natural laws are the laws of physics and chemistry. However, our world also contains non-material entities such as information, will, and consciousness. Persons who hold the opinion that the world can be described in terms of material values only are limiting their scope of awareness.

Using the concept presented here, an attempt will be made for the first time to formulate natural laws for non-material values as well. Since these laws fulfil the same stringent criteria as the natural laws for material values, the conclusions drawn from them are equally informative and valid.

## What is information?

"Information is information, not matter nor energy," is a frequently cited statement made by American mathematician Norbert Wiener. His statement contains a very important element: information is not a material entity.

Imagine the smooth surface of a sandy beach into which I write sentences with my finger. The information content of the sentences can be understood. Then I delete the information by brushing the sand flat again with my hand. Then I write different sentences in the sand, using the same material to present the information as before. Deleting and rewriting the sentences has at no time changed the mass of the sand, although different information has been presented at different times. Thus the information itself has no mass. We could have made the same considerations using the storage medium of a computer.

We therefore conclude information is not a property of matter.

Norbert Wiener told us what information is not. So what is it really? Because information is a non-material entity, its origin cannot be explained on the basis of material processes. What is the initiating factor that results in information existing at all? What moves us to write a letter, a postcard, congratulations, a diary or a file memo? The most important precondition is our own will or the will of the agency directing us. Information is always based on the will of the sender of the information. It is no constant; it can intentionally be expanded, or deformed and destroyed by disturbing influences.

We therefore conclude that information can result only from an act of will or intention.

## The natural laws of information

In order to describe the natural laws of information, a suitable and precise definition is required that will enable us to decide reliably whether an unknown system falls within it or not.

The following definition allows a just allocation: Information is always present if the following five hierarchic levels occur in an observable system: statistics, syntax, semantics, pragmatics, and apobetics.

Statistics (characters): There must be characters for material representation (e.g., letters, magnetizations on a hard disc, DNA base pairs, or sonic spectrum) that can be recorded in the form of statistics. Which individual letters (e.g., a, b, c...z or G, C, A and T) are used? What is the frequency of occurrence of certain letters and words? Claude Elwood Shannon, mathematician and founder of information theory, developed a concept based on this lowest level alone (2) (3).

Syntax (code): The characters are ordered according to certain syntactic rules in a grammar. This second level involves only the character systems themselves (code) and the rules governing how the signs and sign sequences are juxtaposed (grammar, vocabulary), whereby this occurs independently of any interpretation.

Semantics (meaning): Character sequences and syntactic rules are the necessary precondition for presentation of information. However, the decisive aspect of a piece of information to be transmitted is its semantics, the message, the statement, the meaning, and the significance. For example, "GGA" in the code system for living cells represents a glycine molecule.

Pragmatics (will): Information demands action. It is of no consequence here whether the recipient of the information does what the information sender intends, reacts by doing the opposite or does not respond at all. Each case of information transmission is coupled with the intention of the sender to initiate a certain action.

Apobetics (goal): The last and highest level of information is apobetics (purposive aspect, goal aspect). The apobetic aspect of information is most important of all, since it asks about the intention of the sender. Any intelligent information intends something, i.e., pursues a purpose.

# References

1. Werner Gitt. Am Anfang war die Information, 3. überarbeitete und erweiterte Auflage 2002, Hänssler-Verlag, Holzgerlingen.
2. Werner Gitt, 294–311.
3. Claude E. Shannon. "A mathematical theory of communication." *Bell System Technical Journal* 27 (July and October 1948) 379–423 and 623–656.

## Four natural laws about information (NLI)

NLI-1: A material variable can not produce a non-material variable.

    As a matter of general experience, apple tress produce apples, pear trees produce pears and a thistle produces thistle seeds. By the same token, horses bear foals, cows bear calves and women beget human children. In the same way, we observe that a material variable never produces anything non-material. (The term "non-material" is used here instead of immaterial to emphasize the difference from material.)

NLI-2: Information is a non-material fundamental variable.

    The reality in which we live can be separated into two fundamentally different divisions, namely the material world and the non-material world. Matter possesses mass that can be weighed within a gravitational field. By contrast, all non-material variables (e.g., information, consciousness, intelligence, will) have no mass. It must be remembered, however, that matter and energy are required to store and transmit information.

NLI-3: Information cannot be produced in statistical processes.

    Statistical processes are purely physical or chemical processes that run their course without the influence of a guiding intelligence. No information that meets our definition can be produced in such processes.

NLI-4: Information can only be produced by an intelligent sender.

    In contrast to a mechanical sender, an intelligent sender possesses consciousness. It has a will of its own, is creative, thinks independently and acts in a goal-oriented manner.

A number of special natural laws can be derived from general natural law NLI-4:

- NLI-4a: Every code is based on a mutual agreement between sender and receiver.
- NLI-4b: There can be no new information without an intelligent sender.
- NLI-4c: Every information transmission chain can be traced back to an intelligent sender.
- NLI-4d: The assignment of meaning to a set of symbols is a mental process requiring intelligence.

Three remarks of essential importance:

- R1: Technical and biological machines can store, transmit, decode and translate information without understanding the meaning itself. Such cases are covered by NLI-4.
- R2: Information is the non-material basis for all technological and all biological systems.
- R3: A material carrier is required to store information.

The following eight conclusions are drawn with the help of the natural laws of information (NLI).

## Eight far-ranging conclusions

The natural laws NLI-1 to NLI-4 are based on experience. Now we can apply them well-directed and efficiently. This will lead us to eight conclusions answering basic questions.

Since these questions go beyond the limits of scientific actions and thought processes, we require a higher source of information before we can go beyond these limits. This higher source of information is the Bible. In the following texts we will first formulate the conclusion briefly, then explain its reason applying the natural laws of information, then finally provide the Biblical reference confirming the conclusion or even going beyond it.

# 76

# INTELLIGENT INFORMATION

Since all forms of life contain a code (DNA/RNA molecules) and the other laws of information, this clearly falls within the definition of information. We can therefore conclude that there must be an intelligent originator/sender.

## Application of NLI-4

Since there is no provable process in the material world in which information has been produced on its own, this also applies to all information found in living beings. NLI-4 therefore requires here also an intelligent originator who wrote the programs originally.

## Biblical reference:

"In the beginning God created the heavens and the Earth ...And God **said**: Let the Earth put forth vegetation, plants yielding seed, and fruit trees bearing fruit in which is their seed, each according to its kind ...And God **said**: Let the waters bring forth swarms of living creatures, and let birds fly above the Earth across the firmament of the heavens. So God created the great sea monsters and every living creature, ... each according to its kind, and every winged bird according to its kind ...And God **said**: Let the Earth bring forth living creatures according to their kinds, cattle and creeping things and beasts of the Earth according to their kinds" (1).

The word "**said**" is in bold type to emphasize how God created life on Earth by means of His word in the role of an information sender.

## Reference

1. *The Bible.* Gen. 1:1–25.

# 77

# THE OMNISCIENT SENDER

The encoding concept of DNA molecules goes far beyond the capacity of all modern human information technologies. The sender who created the unicellular and multicellular life forms known to us must have been so intelligent, and have been in possession of so much information, that we could, from our point of view, call him infinitely intelligent and omniscient.

(Application of NLI-1, NLI-2 and NLI-4b)

According to NLI-4, there is an intelligent originator at the beginning of every information transmission chain. If this principle is applied consistently to biological information, an intelligent originator is necessary here as well. DNA molecules contain the highest information density known to man (1).

On the basis of NLI-1, the information source cannot be any conceivable processes that take place in matter. Human beings can produce information, but they cannot possibly be the source of biological information. The only alternative left is a sender who acted outside of the world as it is known to us.

Following a lecture at a university, a female student asked, "Who informed God so as to make him capable of programming the DNA molecule?"

**Two explanations are mentally conceivable:**

Explanation a): Let us assume that this God is much more intelligent than we are, but still limited. Let us further assume that He possesses sufficient intelligence or information to enable Him to program all biological systems. The question can then indeed be asked: Who gave Him the required information; who taught Him? In such a case, He would need a higher information provider IP1, i.e., a higher God with knowledge exceeding His. If IP1 knows more than God but is also limited, He would in turn require an information provider IP2, i.e., a God higher than the higher God. This chain of thought could go on forever with IP3, IP4, to IP infinity.

Explanation b): It is more elegant and satisfying to assume only one sender (one originator, one creator, one God). This, however, would require the assumption that He is infinitely intelligent and has an infinite amount of information at His disposal. He must, therefore, be omniscient.

## Which explanation is preferable?

The two explanations are equally logical. Here we are forced to make a decision that cannot be derived from NLI. The following considerations will help us along. In reality, there are only finite and countable amounts. The number of atoms in the universe is an unimaginably large number, but one that could, in principle, be counted. The total numbers of all human beings, ants or grains of wheat are also immense sums, but they are finite sums. The term infinity is commonly used in mathematical abstractions, but reality contains nothing that could be represented by an infinite number.

Therefore, since explanation a) does not pass the plausibility test, the only alternative left is b). This means there is only one sender, who must be infinitely intelligent and omniscient.

## Biblical reference

The Bible teaches us that there is only one God: "I am the first and I am the last, besides me there is no god" (2). What does it mean for God (the sender of the biological information, the creator) to be infinite? It means there is no question He could not answer, and that His knowledge encompasses not only all things of the present and past but also the future as well.

If, however, He has knowledge of all things (even beyond the limits of temporality), then He must Himself be eternal. The Apostle Paul arrives at the same conclusion when he writes, "His eternal power and deity, has been clearly perceived in the things that have been made." (3). The Bible bears witness to God's eternal nature in many different passages (4) (5) (6).

## References

1. Werner Gitt. Am Anfang war die Information, 3. überarbeitete and erweiterte Auflage 2002, Hänssler-Verlag, Holzgerlingen, 311–313.
2. *The Bible*. Isaiah 44:6.
3. *The Bible*. Rom. 1:20.
4. *The Bible*. Psalms 90:2.
5. *The Bible*. Isaiah 40:28.
6. *The Bible*. Dan. 6:27.

# 78

# THE POWERFUL SENDER

The knowledge necessary to program DNA molecules is not sufficient to create life. To make the step from knowledge to practical action, the ability to build all of the necessary biological machines is also required. Life could not have come about without creative power.

Because the sender encoded the information in the DNA molecules in such an ingenious manner, and because he must have constructed the complex biological machines that decode the information and carry out all of the biosynthetic processes and formed all of the structural details and abilities of the living organisms, we can conclude that the sender must be powerful.

On the basis of natural laws (NLI-1, NLI-2 and NLI-4b), we determined that the sender of the information in the DNA must be omniscient. The question here is how great his power actually is. "Power" here is understood to include everything we mean to describe with the terms ability, strength, effectiveness, and creativity. This kind of power is necessary to make every living thing.

## Biblical reference

We have no quantitative notion of the extent of this enormous power, but the Bible tells us of its true extent by presenting the sender behind it as all-powerful: "I am the Alpha and the Omega, who is and who was and who is to come, ...the Almighty"(1). "For with God nothing will be impossible" (2).

## References
1. *The Bible*. Rev. 1:8.
2. *The Bible*. Luke 1:37.

# 79

# THE NON-MATERIAL SENDER

Because information is, by its very nature, a non-material value, it cannot have been derived from a material value. We may thus conclude that the sender must be non-material (spiritual) by his very nature.

## Application of NLI-1 and NLI-2

Information is a non-material value, for which reason its source must also be non-material. We can therefore conclude that the sender must be non-material by his very nature or that he must at least include a non-material component.

## Biblical reference

We learn from the Bible that God is spirit (1), that the material world is subject to Him (2), that He Himself is non-material (3), that He speaks and it comes to be (4).

## References

1. *The Bible*. John 4:24.
2. *The Bible*. Luke 8:28.
3. *The Bible*. Kings 19:11–13.
4. *The Bible*. Psalms 33:9.

# 80

# REBUTTAL OF MATERIALISM

Human beings are capable of creating information. Since this information is of a non-material nature, they cannot come from our material component (body). We can thus conclude that human beings must have a non-material component (soul, spirit).

## Application of NLI-1, NLI-2

The thought processes in the fields of evolutionary and molecular biology are based exclusively on material concepts. This reductionism (allowing material explanations only) has actually been elevated to the status of a working principle. Based on the principles of information, materialism can be refuted as follows.

We all are capable of producing new information. We can record thoughts in letters, essays and books or carry on creative discussions. In so doing, we produce a non-material item, namely information. The fact that a material carrier is required to store and pass on the information does not alter the essential nature of information.

This enables us to draw a very important conclusion, namely that we must possess a non-material component in addition to our material bodies. The philosophy of materialism, most prominently featured in Marxism-Leninism and Communism, is thus refuted based on the natural laws of information.

## Biblical reference

The Bible affirms that man is not purely material (1). The body is the material part of a human being, but the soul and spirit are non-material (2).

## References

1. *The Bible*. Gen. 2:7.
2. *The Bible*. 1 Thes. 5:3.

# 81

# REBUTTAL OF THE BIG BANG THEORY

The claim that the universe emerged solely from a singularity (scientific materialism) contradicts the non-material magnitude of information. Therefore, a big bang is excluded as the sole cause of the origin of the universe.

(Application of NLI-2)

According to the principal proponents of the big bang theory, everything we perceive, observe and measure in our world had its sole origin in matter and energy without any further added components. This hypothesis can be refuted by applying the natural laws of information just as the concept of a perpetuum mobile can be refuted.

Our world contains an abundance of information in the cells of all living things. According to law NLI-1, information is a non-material magnitude and can therefore not have originated from matter and energy. The big bang theory is therefore false.

The proponents of the theory of evolution see it as a universal principle. It is said to form a chain in which each link is irreplaceable: Big bang – cosmological evolution – geological evolution – biological evolution. If one link does not hold, the strength of the entire chain is lost. The refutation of the big bang theory breaks the very first link in the chain.

This conclusion can also be formulated as follows: No big bang system is conceivable that could result in the origin of information and life.

## Biblical reference

The Bible teaches us that this world did not come about as a result of a process lasting thousands of millions of years, but was rather created by Almighty God. "For in six days the Lord made heaven and Earth and all that is in them" (1) (2).

## References
1. *The Bible.* Exodus 20:11.
2. *The Bible.* Job 38:1-41.

# 82

# ABIOGENESIS AND MACRO EVOLUTION

Because the basic component of all life is information, which cannot originate from matter and energy, an intelligent sender is required to create the information in the DNA molecules. Since, however, all theories of chemical and biological evolution require that the information originate solely from matter and energy without an intelligent sender, we may conclude that all of these theories and concepts of abiogenesis (creation of life from non-life) and macro evolution must be false.
 (Application of NLI-1, NLI-2, NLI-4b, NLI-4d)
Life, that which is animate, is a non-material value that could not have originated by matter. Purely material processes cannot produce life on Earth, nor can they do so elsewhere in the universe. The claim that life could have come solely by way of material processes, as soon as the basic conditions are met, contradicts empirical experience. Biologist William Dembski has made a similar statement (1).

## Impermissible reductionism

The theory of evolution attempts to explain life solely on a physicochemical basis (reductionism). Reductionists would even prefer a fluid transition from the inanimate to the animate. We can draw a very basic and far-reaching conclusion based on the natural laws of information:

The concept of abiogenesis and macro evolution, the pathway leading from inanimate matter to human beings, is false. Information is a basic and indispensible component for all living systems. All intelligent information, including the one in animate systems, requires a mental creator. In the light of the natural laws of information the evolutionary system is a "perpetuum mobile of information."

## Where do we find the sender of the information in the DNA molecules?

The sender of the information in living beings is not readily apparent. Does this justify the conclusion that the information must somehow have been generated by molecular biological means?

With regard to the abundance of information contained in the hieroglyphs of Egypt, none of the stones show us anything of the sender. We find there only traces he chiselled into the stone. Nonetheless, no one could possibly contend that this information originated without a sender and without a mental concept.

When two computers are interconnected so as to exchange information and initiate certain processes, the sender is also not evident at all. However, all the information involved was at some time conceived by a programmer or programmers.

Just as a computer transmits information to another computer, the information in the DNA molecules is transferred to RNA molecules. Each living cell contains enormously complex bio machinery in which the programmed commands are executed in an ingenious manner. Despite the fact that we cannot see the sender of the information, ignoring him would constitute impermissible reductionism.

## Parametric optimizations in the biological information

It is no wonder that the programs of the sender of biological information are much more ingenious than all human programs. We are dealing here with a sender of infinite intelligence. The program of the creator is conceived so ingeniously that extensive adaptations to new conditions are also possible. In biology, such processes are termed micro evolution.

The natural laws of information exclude the possibility of macro evolution as required by the theory of evolution. On the other hand, variations, often involving extensive adaptations within a species are explainable based on the ingenious program of the creator.

## Biblical reference

The story of creation as told in the Bible repeats emphatically, nine times, that all plants and animals were created "each according to its kind" (2).

## References

1. William A. Dembski. *The Design Revolution.* (Inter Varsity Press, 2004): 157.
2. *The Bible.* Gen. 1:20–25.

# 83

# OLD AND NEW PROOFS FOR THE EXISTENCE OF GOD

The causal proof of God's existence formulated by Aristotle (384–322 B.C.) assumes that the series of causal movers cannot be infinite, so that there must be a prime mover (prima causa). In the ontological proof of God's existence, Anselm of Canterbury (1033–1109) draws his conclusion by moving from the logical, terminological level to the level of being. The teleological proof of God's existence of Thomas of Aquinas (1225–1274) states that the ordered and obviously planned nature of the world must have an external cause. There are a number of variants on the cosmological proof of God's existence. The earliest formulation argues that the universe requires a causal agent that must lie outside it. More recent proofs of God's existence can be derived from the natural information in the universe and the prophetic information in the Bible.

Proofs of God's existence have always been a focus of both strong support and equally adamant criticism (1). Most of the critical commentators refer to Immanuel Kant, considered the main destroyer of all such proofs. Together with the poet Gotthold Ephraim Lessing, Kant has become the very embodiment of the Enlightenment, his definition of which was "humanity breaking the bonds of immaturity it had forged for itself." The two of them were known as the "Twin stars of the Enlightenment," i.e., of the movement that claimed that the Bible was implausible.

Kant viewed our perception capacity as very limited, despite our brain which is constantly asking about the meaning of life, the soul, and God. The Bible says that we can indeed know God (2). Also, "It is the Spirit himself bearing witness with our spirit that we (Christians) are children of God" (3). The clearest revelation of all comes from Jesus Christ himself: "He who has seen me has seen the Father" (4).

In the Bible, God leads us to the correct view. He explains that we can conclude that God exists based on the created works with the help of our mind: "For the invisible of him from the creation of the world are clearly seen, being understood by the things that are made, even his eternal power and Godhead; so that they are without excuse; for although they knew God they did not honour him as God or give thanks to him" (5).

The formulation "they knew God" is a highly significant statement. It declares that God has revealed Himself outside of the Bible as well. While proofs of God's existence alone may not lead to faith, they do have an important function: They refute atheism and are a suitable means of reducing, or even eliminating, some obstacles to faith.

## The proof of God's existence based on the natural laws of information (6)

On the basis of the natural laws of information (NLI), we know that the enormous amounts of information in the cells of all living things require an intelligent originator. Compared to the historical proofs of God's existence with their mainly philosophical arguments, what we have here is proof based on natural laws of the existence of an intelligent sender and thus for the existence of a God. Kant knew nothing about the existence of the genetic information, so that modern proofs of God's existence can not refer back to Kant, who lived over 200 years ago and was aware of only a small part of the knowledge that the natural sciences have uncovered since that time.

## The prophetic-mathematical proof of God's existence (7)

The notion is widespread today that the Bible is just a book like any other book. Is it true that it is just a collection of writings by people who have thought about God and the world down through the ages?

The Bible contains 3,268 prophecies that have already been fulfilled (8). This is true for no other book in world history (9). This provides us with a unique criterion for testing the truth of this work. Is it possible that people could have made that many precise predictions distributed over a period of 1,500 years? Were the prophecies fulfilled by chance, or was it only possible because God is the author of the Bible and is able to make prophecies because of His omniscience that can then be checked for accuracy as history unfolds?

The probability that 3,268 prophecies would be fulfilled by chance is practically zero. The results of the relevant mathematical calculations are so gigantic, even hyper astronomic, that our mental capacity and imagination cannot conceive of the figures involved. Assuming that all prophecies have a fifty percent probability, the degree of probability resulting from the calculations is an unimaginable $1.7 \times 10^{-984}$.

## Four direct conclusions

a) It is not conceivable that all of the biblical prophecies that have been fulfilled have come true by chance. This critical objection can be eliminated statistically.
b) Since the prophecies as a whole could not have been fulfilled by chance, an omnipotent and omniscient

God must have foretold these things and then brought them to pass by virtue of His omnipotence.

c) Since the prophecies can only be fulfilled by an omnipotent and omniscient God, our observations have led to a prophetic-mathematical proof of God's existence. This could also be expressed by saying that the idea of atheism has been refuted.

d) Since we were concerned here with the prophecies in the Bible, the God named under b) is no other than the living God of the Bible, who revealed himself through normal people and came to us personally in Jesus Christ.

Two indirect conclusions

e) Of the total of 6,000 prophecies in the Bible, 3,268 have already been fulfilled. Many prophetic statements (in particular in the Book of Revelation) refer to the return of Jesus and the end of world history and have not been fulfilled as yet. We can, however, draw the indirect conclusion that these prophecies will also be fulfilled in due time, exactly as described.

f) If we have established that large parts of the Bible were inspired by the infinitely intelligent and omnipotent creator of the universe, it is actually inescapable that the entire Bible, including its statements about creation, must be true.

Two overall conclusions

g) The prophetic-mathematical proof confirmed the existence of an omniscient and omnipotent God, who must be identical with the God of the Bible.
The Bible was authored by God, and it is true.

## Conclusion

None of the proofs of God's existence cited from the past refers to a specific God. They are all of such a general nature that any religion could make use of them. The prophetic-mathematical proof of God's existence, on the other hand, clearly refers to the God of the Bible and His son Jesus Christ.

## References:

1. Alister McGrath. Der Atheismus-Wahn, Gerth Medien, 2007.
2. *The Bible*. Rom. 1:19.
3. *The Bible*. Rom. 8:16.
4. *The Bible*. John 14:9.
5. *The Bible*. Rom. 1:20–21.
6. Werner Gitt. Am Anfang war die Information, Hänssler-Verlag, Holzgerlingen, 3. überarbeitete and erweiterte Auflage, 2002.
7. Werner Gitt. So steht's geschrieben, 7. stark erweiterte und überarbeitete Auflage, Christliche Literatur-Verbreitung, Bielefeld, 2008.
8. Finis Jennings Dake. *Dake's Annotated Reference Bible*. Lawrence Ville, Georgia, USA, 1961.
9. Werner Gitt. Und die anderen Religionen? Christliche Literatur-Verbreitung, 1991.

# HUMANS AND CULTURE:

One impressive argument against the theory of evolution is man himself. Is this what an accidental product of an endlessly long developmental process would look like?

It is not only the outer geometry of the body structure, but also the human spirit, human language, the human eye, human hands, etc. – everything so perfectly and uniquely formed that an accidental and random development cannot explain it. Then of course there is also the amazing human faculty that enables us to be creative, to create new things.

Another question needs to be asked as well. If the human spirit is really a product of material processes, how can it be that man can conceive things that go beyond his material existence and become conscious of him? Philosopher René Descartes defined himself in terms of the ability to think. "I think, therefore I am," was his conclusion. Can the human psyche be explained as a material phenomenon? Can extrasensory and spiritual experiences be explained in purely naturalistic terms?

Human historical traditions, geology and archaeological discoveries include evidence that contradicts the notion of an extremely old Earth. Based on the scenario of a worldwide flood, the development of the geological formations can be readily explained with the model of a young Earth and a brief history of mankind.

# 84

# REPORTS OF THE FLOOD

In older human cultures and on all five continents, historical traditions tell of a major flood event. Accordingly, geological strata all over the world indicate several continental flood catastrophes that can best be interpreted as post-flooding events following a single, gigantic, worldwide flood.

A diluvian flood is reported in seventy-seven different cultures spread over the entire globe, and the rescue came in the form of a ship in seventy-two of them. It would be surprising if such a catastrophe were not remembered in the historical traditions of the different peoples. Some circumstances that support these flood reports are:

## Mass fossil graves

Mass fossil graves of enormous extent are found on all continents of the Earth. For instance, the Old Red Sandstone in Scotland (160 km from the Orkneys) abounds with petrified fish that died a violent death. It is estimated that the Karroo formation ($518,000 km^2$, an area of extensive rock strata in South Africa) contains the skeletons of approximately 800 billion animals, mainly amphibians and reptiles (1).

## Extensive coal strata

The extent and distribution of coal deposits worldwide cannot be explained by slow processes. A peat layer approximately

50 m thick is required to form a coal seam one m thick. A coal seam 10 m thick would require collection of a layer of plant material about 500 m thick. These facts are most readily explained as the result of a gigantic flood catastrophe in which floating material drifted together and was then covered with sand and mud from the interior of the country (2). Since practically no roots had grown into the lower layers, the coal seams are evidence of a rapid deposition process.

## Gigantic erosive floods

Ice age cataclysmic floods, for example the Missoula floods in the north-western US, cut valleys hundreds of metres deep into hard rock. The geological debates lasted for decades until it finally became clear that these geographic findings could only be grasped in terms of a catastrophe (3). In *Science's* March 29, 2002 edition, Victor R. Baker reported how many geologists had long ignored the effects of superfloods (4). It was generally assumed that the great majority of canyons and valleys had been formed by the gradual forces of wind and water over periods of thousands and millions of years. However, the new perceptions lead to a gradual rethinking (5).

## Origins of the Grand Canyon

Many geologists realize that the 28 km-long Grand Canyon could not possibly have been eroded by the Colorado River. It is easy to imagine a huge mass of water being dammed up behind the Kaibab Upwarp plateau towards the end of the post-flood events. Ice age rains may then have filled this lake to overflowing, whereupon the onrush of flood waters with their freight of rocks and boulders were able to carve out the gigantic valley of the Grand Canyon within a brief period (6).

## Continent-wide sand distribution

Much of the sand found today has been transported over great distances. Gravel and sand covering the Sahara over an area larger than 1,000 x 1,000 km was transported from the sea into the interior of the country and evenly distributed there (7). The silicate sands of Florida (USA) come from the Appalachians and were thus transported over a distance exceeding 700 km (8). The quartz gravel in North Dakota (USA) comes from the area around British Columbia (Canada) and must also have been transported farther than 700 km (9).

## Noah's Ark

Of the seventy-two flood reports including a ship, the report of the biblical ark built by Noah is clearly the most sensible and logical:

The proportions of the biblical ark result in optimum floating stability, comparable to that of a modern container ship. The amount of material needed to construct the ark, with a width to height ratio of 0.5, is the lowest (10). The interior of the ark was sufficiently large enough to accommodate all animal species requiring protection from the flood, including food reserves sufficient to last for one year (11).

## Continental drift (plate tectonics)

It can be assumed that all of the present-day continents were joined together immediately after the flood. Of the land animals that had survived in the ark, some spread rapidly throughout the continent and others found habitats in various regions therein. Rapid continental drift in the ensuing period put an end to this distribution process, so that marsupials, for example, are now found almost exclusively in Australia.

A gradual continental drift is still observed today, whereby it is quite conceivable that the rate at which this process takes place was much higher during the flood and in the postdiluvian centuries. Geophysicist John R. Baumgardner developed a computer simulation of the process modelled on such a scenario (12). To reach their present positions, Africa and America, for example, must have been moving apart at a rate of 12 cm/h during 500 years.

## High-pressure minerals in subduction zones

The rapid return of mineral types from subduction zones (tectonic plate submersion zones) demonstrates that rapid shifts in the Earth's crust are still possible today. The presence of high-pressure minerals, e.g., in the Dora Maira massif in the western Alps, shows that such mineral types can rise to the surface quite rapidly. Geologists Frisch and Meschede describe,, "This convoluted gliding and shifting process ... can raise deep-lying minerals rapidly to near-surface strata. Destruction of these high-pressure minerals takes place when the rocks rise gradually or as a result of the influx of water during tectonic processes while they are rising. High-pressure minerals are only preserved if they rise rapidly and ... are cooled quickly" (13).

## References

1. David C.C. Watson. Weltschöpfung und Urgeschichte, 166–167.
2. Joachim Scheven. Karbonstudien, Neues Licht auf das Alter der Erde, Hänssler, 1986.
3. Stephen. J. Gould. Der Daumen des Panda, Suhrkamp, 2. Aufl., 2008, 204–214.
4. Victor R. Baker. *Science 295* (29 März 2002): 2379–2380.
5. Alexander und Edith Tollmann. Und die Sintflut gab es doch. Vom Mythos zur historischen Wahrheit, Droemer Knaur, München 1993.
6. John D. Morris. *Geology*. Master Books, 69. (deutsche Fassung: factum August 2007, 22–30).

7. H. Füchtbauer und G. Müller. Sedimente und Sedimentgesteine II, 1977, 3. Auflage, Stuttgart.
8. Carl R. Froede Jr. CRSQ 42 (März 2006): 229.
9. Michael J. Oard. CRSQ 44, Spring 2008, 264.
10. Werner Gitt. The sonderbarste Schiff der Weltgeschichte, Fundamentum 3/2000, 36–81.
11. Fred Hartmann und Reinhard Junker, Passten alle Tiere in die Arche Noah? Wort und Wissen, Diskussionsbeitrag 4/90, http://www.wort-und-wissen.de/index2.php?artikel=disk/d90/4/d90-4.html
12. John R. Baumgardner. "Runaway subduction as the driving mechanism for the Genesis Flood." Proceedings of the Third International Conference on Creationism, 1994, Pittsburgh, Penn., USA, 63–75.
13. Wolfgang Frisch und Martin Meschede. Plattentektonik, Kontinentalverschiebung und Gebirgsbildung, 2007, Wissenschaftliche Buchgesellschaft, Darmstadt, 117–118.

# 85

# THE AGE OF HUMANITY

Most experts believe that humans have existed for about two millions years. This would mean, however, that the rate of population growth must have been practically zero up until the modern era. Comparisons with similar cultures present today show that this scenario is unrealistic. In addition, an analysis of the remains of Stone Age people tells us that, even though they had enough to eat, their populations never numbered in the millions. The empirical demographic and volumetric data and the estimates of what they have left behind allow at most some thousand years of prehistory. Six aspects which contradict the hypothetical two-million-year history of humankind:

1) The lack of population growth
2) Cultural and technical stagnation
3) The small numbers of stone tools found
4) The low levels of settlement stability and relatively small numbers of settlement sites
5) The brevity of cave habitation periods
6) The lack of graves

## 1) The lack of population growth

Assuming poor to catastrophic living conditions for early humans, and assuming a very low level of annual population growth of 0.1%, there should have been eight million Stone Age humans after only 15,000 years. Even under the worst possible conditions, the human population of the Earth would have to become about as large as it is today after 23,000 years at most. The marks which they have left behind indicate that the living conditions (nutritional situation and health) were in most cases quite good, so that one would have to assume a more rapid growth of population (1).

## 2) Cultural and technical stagnation

Cultural and technical development stagnated almost completely throughout the entire Stone Age. The reason given for this is that early humans were supposedly mentally underdeveloped. The archaeological remains, however, tell a different story. They tell us that both the Neanderthals and Homo erectus had skills and behavioural features that match those of modern humans (2) (3). The archaeologist Robin Dennell describes in his writing the remarkable depth of planning, refined sense of design and patient woodcutting that went into making weapons (4). Up to date, these qualities had been ascribed to modern humans only.

## 3) The small numbers of stone tools found

Conventional science estimates the age of the remains of the earliest true humans at about two million years. It is assumed that these people lived as hunters in a Stone Age culture until 10,000 years ago. However, remains from these people have not been found in significant numbers at all. This is particularly remarkable with reference to the stone tools, since they are preserved relatively well over time. Counting the tools found from earlier times and comparing these counts with the numbers of tools made by hunting cultures of today reveals that the number of stone tools found is much too small.

Even assuming that, for example, a population of only 1,000 persons had lived in Germany over a period of 800,000 years, we should be able to find several billion stone tools. Realistically, however, it must be assumed that the population of Europe was several millions, at least in some phases. Compared to the billions of stone tools they would have to have left behind, the actual stone tool findings are meagre indeed (5).

## 4) The low levels of settlement stability and relatively small number of settlement sites

The number of settlement sites of Stone Age humans is also much lower than it should be. For example, consider the situation in Bohemia during the Magdalenian period, purportedly from 11,500 to 15,000 years ago. An estimated population of 350 persons is said to have lived in the area, distributed among fourteen groups. They moved camp several times a year. In a period of 3,500 years, these fourteen groups alone must have left behind between 87,500 and 245,000 campsites. Only fifteen have been found to date. Even allowing that only a very small proportion of the settlement sites may have lasted down to the present, this number is simply much too small for the length of time in question. It is, by the way, also highly unrealistic to assume that a population of 350 persons could survive for more than 3,500 years without increasing (6).

## 5) The brevity of cave habitation periods

Contrary to what one might think, the cave habitation periods were brief in each instance. This is also evident from the few remains found in the caves of south western Germany. For instance, only three sites were found in the Eselsburg Valley from a period claimed to be 25,000 years long, and these sites had only been inhabited for short periods. If humanity had lived at least occasionally in caves for over one million years with only a minimum increase in population during that time, it would have to be assumed that a very large number of caves would have been inhabited for periods extending to thousands of years (7).

## 6) The lack of graves

Even assuming a minimum population density of only three inhabitants per km$^2$, this would result in a grave density of 0.15 graves per m$^2$ over a period of 1.5 millions years (that is one grave every 2.6 m). Of course an individual grave was not dug for each person. Nonetheless, if the history of mankind really had lasted two million years the continents should be literally covered with graves.

## References

1. Michael Brandt. Wie alt ist die Menschheit?, Hänssler-Verlag, 2006, 67–86.
2. Hartmut Thieme in einem Interview in Spektrum der Wissenschaften, Oktober 2004, 48–50, Jagdwaffen and -strategien des Homo erectus.
3. Junker and Scherer. Evolution, ein kritisches Lehrbuch, 2006, 283–286.
4. Robin Dennell. "The world's oldest spears." *Nature* 385 (27 Februar 1997): 767–768.
5. Michael Brandt. Wie alt ist die Menschheit?, Hänssler-Verlag, 2006, 95–123.
6. Robin Dennell, 125–129.
7. Robin Dennell, 137–140.

# 86

# NEANDERTHALS AND AUSTRALOMORPHS

The descent of man from apelike ancestors has still not been proved. Not a single indisputable example of the fantastic intermediate forms (ape becoming human with upright gait) published in the media has ever been found. The most famous prehistoric human, numerous examples of which have been unearthed, is the Neanderthal. The Neanderthals were by no means primitive. Quite the contrary, their average braincase volume was larger than that of modern humans. Whether and to what extent modern humans are related to the Neanderthals is a matter of controversy. At any rate, it seems clear they could not possibly be the "missing link" between apes and humans. The hypothesis generally supported today is that Neanderthals, chimpanzees and modern humans share a common ancestor. However, not a trace of these hypothetical ancestors has ever been found. The Australomorphs also do not appear to be the missing links.

The extinct anthropoid ape genus Australopithecus, as well as several other similar genera (known collectively as "Australomorphs"), is under consideration by evolutionary theorists as potential human ancestors. However, all of these forms show characteristic features that are irreconcilable with the transition form status. The same applies to the recently discovered genera Orrorin, Kenyanthropus and Sahelanthropus.

The known fossils cannot be arranged in an indisputable line leading to humans. Each species involved has features that contradict the assumed lineages. The Australomorphs do not fit the role of either connecting links between apelike species and Neanderthals or ancestors of modern humans (1).

## Conclusion

It is not sufficient to emphasize individual features that would appear to support a transitional role for Australopithecus (and other genera) between anthropoid apes and humans. The overall spectrum of characteristics is the decisive factor. The macro evolutionary hypothesis does not require the evolution of individual characteristics, but rather the whole species. To qualify as a transitional form, the spectrum of characteristics as a whole should correspond to a transitional form, at least approximately. This is clearly not the case with Australopithecus.

Due to the overall unique character of their spectrum of characteristics, the australomorphs could be considered an independent extinct basic type with no ancestral relationship to humans.

Ramapithecus, once touted as the first human-like animal and early ancestor of humans, is now considered to be more likely a relative of the Asian anthropoid ape species orang-utan (2).

## References

1. Sigrid Hartwig-Scherer. 18.06.2007, http://www.genesisnet.info/?Sprache=de&Artikel=43622&l=1.
2. Sigrid Hartwig-Scherer. Ramapithecus, Vorfahr des Menschen?, Zeitjournal-Verlag, 1989, 47.

# 87

# THE HUMAN AND CHIMPANZEE GENOMES

The difference between the genomes of humans and chimpanzees has been claimed in the past to be 1.5 to 2%. Molecular biologist Roy J. Britten has determined that, when the insertions and deletions are included in the calculation, the difference is nearly 5%. This means that at least seventy-five million "correct" mutations would have been necessary to make a modern human and a chimpanzee from a common ancestor. Even if one advantageous mutation per year had occurred in these populations, a total of seventy-five million years would have been necessary, whereas the evolution of humanity is supposed to have taken only two million years. According to calculations by genetic pioneer J.B.S. Haldane, a realistic estimation of the time that would have been required for this process is at least 2.5 billion years.

The scientific consensus today is that the difference between the genome of humans and chimpanzees amounts to 1.5 to 2%, purportedly supporting a relationship between apes and humans. However, Britten determined that the difference is nearly 5% when the insertions and deletions are included in the calculation (1) (2). It is also possible that even a greater difference may be discovered, since only a small fraction of the genome has been compared to date. Of the total of three billion base pairs in the human genome, only about one million had been compared by the year 2008.

# Differences between the human and chimpanzee genomes

1) Humans have twenty-three chromosome pairs, chimpanzees twenty-four.
2) There are special sequences at the end of each chromosome called telomeres. Apes have about twenty-three kilo basepairs, humans only ten.
3) Whereas eighteen chromosome pairs are practically identical, the genes and markings are in a different order in chromosomes four, nine, and twelve.
4) The Y chromosome has a different size and many markings that do not match.
5) Chromosome twenty-one contains large regions that are completely different.
6) The chimpanzee genome is 11.5% larger than the human genome.

As we said above, the chimpanzee genome is 11.5% larger than the human genome (3). How could there be a difference of only one percent if the genome of the chimp contains 11.5% more? This is not possible. The 11.5% are simply ignored. In fact we do not yet know the difference between the human genome and the genome of the chimp (4).

A book we recommend on this subject is *Genetic Entropy & the Mystery of the Genome*, published by geneticist John C. Sanford in 2005. Sanford shows that the genome loses more and more information over time until the species goes extinct.

# Haldane's dilemma

When a useful mutation occurs in a population, as many copies of it as possible must be distributed so that evolution can continue. In other words, the individuals that do not yet contain this mutation must be replaced. The rate at which this process can take place, however, is limited. One of the main limiting factors is the propagation rate of the given species. For a hominid species with a generation interval of twenty years and a low reproductive rate per individual, mutations spread very slowly through the population (5).

John B.S. Haldane (1892-1964) is one of the three founders of the modern science of population genetics. He assumed for a rough calculation a population of 100,000 ancestors in which one male and one female underwent a mutation at the same time that was so beneficial that they outlived all the others, which is also highly unlikely. All the rest (the other 99,998) of the population died out and the surviving pair multiplied, eventually replenishing the entire population. This process would have to be repeated over the course of ten million years in each generation (i.e., every twenty years) to introduce 500,000 (10,000,000/20) advantageous and perfectly-adapted mutations into the population. These 500,000 mutations would then amount to only 0.02% of the necessary 5%. If more realistic rates of fitness/selection and population renewal are assumed, even 2.5 billion years will not be nearly enough.

Haldane's dilemma was still a topic of discussion in scientific journals back in 1960, but the subject has been ignored since that time (6). This may be because mathematical modelling of such processes in population genetics is extremely complex. Research in the field today concentrates primarily on determining the number of advantageous mutations that can be determined to have actually occurred. Important elements that would enable us to continue with such calculations are still missing.

In 1992, the well-known evolutionary geneticist George C. Williams remarked, "The time has come for renewed discussion and experimental attack on Haldane's dilemma" (7). The appeal apparently produced no echo among his colleagues. Walter ReMine published a large study in 1993 in which he investigated the matter in detail (8). He continued work in this field, refined his arguments and addressed attempts by evolutionists who would like to obfuscate the matter. Unfortunately, no serious dispute of the matter has resulted to date. ReMine reminds us that Haldane's dilemma has never been solved, but only hushed up, misrepresented and prematurely dismissed (9).

## References

1. Roy John Britten. "Divergence between samples of chimpanzee and human DNA sequences is 5% counting indels." *Proc. Nat. Acad. Sci.*, 99 (2002): 13633–13635.
2. David A. DeWitt. "98% Chimp/human DNA similarity? Not any more." *Technical Journal* 17/1 (2003): 8–10.
3. CRSQ 45/4, (2009): 242-243.
4. Chimpanzee Sequencing and Analysis, Consortium (CASC). 2005. Initial sequence of the chimpanzee genome and comparison with the human genome. *Nature*. 437:69–87.
5. John Burdon Sanderson Haldane. "The cost of natural selection." *Journal of Genetics* 55, (1957): 511–524.
6. Don Batten. "Haldane's Dilemma has not been solved." *Technical Journal* 19/1 (2005): 20–21.
7. George Christopher Williams. *Natural Selection: Domains, Levels and Challenges*. (New York: Oxford University Press, 1992): 143–144.
8. Walter J. ReMine. *The Biotic Message*. (St. Paul Science, St. Paul, MN, 1993.)
9. Walter J. ReMine. "Cost theory and the cost of substitution - a clarification." *Technical Journal* 19/1 (2005): 113-125.

# 88

# UPRIGHT GAIT

The upright gait of humans involves concurrent occurrence of the following anatomic features: stretched knee and hip joint, cervical spine joined to the head at a point centred beneath it (instead of at the back as in apes), flat face, improved organ of equilibrium, straight back, foot with raised instep, a strong big toe and brain functions facilitating an upright gait. Each of these features would have required the simultaneous occurrence of several thousand correct and perfectly-adapted mutations in the genome, a sheer unthinkable scenario.

## Unique features of human upright gait

The human foot is formed to the best advantage for the upright gait. It forms a flat arch between the ball and heel. This facilitates better balancing on uneven terrain. The foot has twenty-six bones as well as many muscles and tendons that give it the flexibility it needs to make walking function well. Thanks to the foot's arched instep, it can absorb the impact of walking and running. Apes, on the other hand, have hand-like feet that make it easier to grasp branches, but harder to walk.

The big toe of the human foot is particularly strong. It lies parallel to the other toes. At every step, the big toe provides the final abetting. The big toe must be particularly strong to keep the body under control when walking. In apes, on the other hand, the big toe is splayed, enabling it to grasp and hold a branch readily.

The human knee joint makes it possible to extend the legs until they are straight. In the upright position, the knee joint assumes an extended and locked position that relieves the muscles while in standing position. Apes cannot extend their knee joints completely, forcing them to walk with bent legs, a very tiring affair. Scott F. Dye, M.D. wrote the following passage on the uniqueness of the human knee, "Despite the general similarity of the knee among all tetrapods (land vertebrates), none of them provides an ideal model for the human knee" (1).

Human legs are about half as long as the entire body, making it possible to walk or run long distances. The legs of apes, on the other hand, make up only about one-third of the length of the body, so that these animals tire more quickly from walking. A chimpanzee is unable to stretch its legs out straight when standing and finds the upright position very strenuous. Its face is then turned upwards so that it must bend over to look ahead. In apes, the spine is joined to the back of the head, in humans to the bottom of the head. This means apes can look straight ahead readily when they move on all fours, but not when they stand upright. A small human child crawling on all fours has to expend considerable effort to hold its head up in order to look straight ahead.

The hip joints of the human body make it possible to move the femoral bones into a vertical position. This is not possible for apes. The human femur (thigh bone) is constructed so as to allow the knees and feet to be close together when standing. The position of the feet near the centre of gravity of a standing person makes for the greatest possible degree of stability for walking and running. While we walk or run, the body is carried alternately by one foot at a time and would therefore tend to tip over if the centre of gravity were too far outside the load bearing foot. The femurs of apes, on the other hand, are straight, so that the knees are farther apart. That is the reason apes sway from side to side so much when they attempt to walk upright.

The straight back of humans positions the head directly above the hips when a person is standing. Apes have a bent back and must therefore use their hands for support to keep themselves from falling over. The human spine is slightly S-shaped, whereas the spine in apes has a shape of a C. The ape spine is relaxed when the animal walks on all fours. In humans, the spine is relaxed when we walk upright. Any transitional forms would have had to bear an unfavourable load. In this connection, it is interesting to note that present aboriginal groups have a healthy upright gait. The only creatures that in some cases go through life with a bent posture are civilized urban humans.

The flat face of humans makes it possible for us to see what is right in front of us. The chimpanzee, on the other hand, has eyes set farther back in its head and a protruding chin, making it unable to see an obstacle right in front when walking upright. The ape's head is lower when he walks on all fours, thus enabling it to see the obstacles.

The organ of balance in the human ear is designed to master the vertical spatial dimension in particular. In apes, on the other hand, the capacity to balance in the anterior vertical dimension is much less pronounced (2). When walking on all fours, apes are normally already balanced by the four contact points. Apes cannot walk on their toes or stand on one leg.

Facial expression is an important part of human communication. We may not always be conscious of the fact, but we actually keep a constant eye on the facial expressions of the persons within our field of view. We try to guess their thoughts and reactions. Many of our own reactions are influenced by the facial expressions of others. For example, when we see someone with a sad face, we ask why. Apes have relatively few facial muscles and can only change their facial expression within narrow limits.

The human vocal organ is conceived for the exchange of information using a language. Apes produce vocalizations using a different construction. The human larynx lies deeper in the throat, giving the tongue more room to move. This makes it possible to produce more vowels. In apes, the larynx lies much higher, making production of precise sounds impossible. The human oral cavity is also acoustically advantageous.

The ability to speak depends on a corresponding part of the brain that controls the muscles used in speech and processes the signals received by the aural sense so as to make them comprehensible. This brain region is lacking in apes.

The human brain is much larger than the ape brain. The human brain contains about 100 billion neurones, each of which has about 1,000 interconnections to other neurones. Counting the number of connections to the cerebral cortex at the rate of one connection per second would take 3.2 million years.

The ability to think is what makes a human being human. He is self-aware and creative. The human brain has the unique ability to recognize beauty. The left half of the brain contains the centre for processing and producing language and the right half contains the centre for making and perceiving music. No other creature has an "ear for music."

# References

1. Scott F. Dye, M.D. "An evolutionary perspective of the knee." *Journal of bone and joint Surgery* 69A (1987): 976–983.
2. Labyrinth und aufrechter Gang, factum Mai 1995, 17–21.

# 89

# THE HUMAN EYE

The retina of the human eye contains 126 million pixels (image points). An average digital camera has only six million pixels. The signals from the eye's pixels are first compressed by special nerve cells, and then transmitted through about twelve million neural fibres to the brain. Each individual neural fibre must reach a certain location in the brain for the image to be produced correctly there. This allocation of the neural fibres cannot possibly have been the result of a step-by-step, chance process. Another difficulty is that the fibres must be crossed, fanned out and directed to various specific areas on their way to the brain.

The image we see is transformed into electrical signals on the retina of the human eye. An enormous number of nerve fibres conduct the signals from the retina to various layers of the brain. Perception of the image then takes place in the brain.

How is it possible for every single one of these millions of nerve fibres to be directed from the retina to the right location in the brain during the growth of a creature? Could it be that each individual nerve fibre gradually ends up at the right place by "trial and error"?

The fovea centralis (the location in the eye with the sharpest focus) contains about 15,000 pixels. The signals from them are collected in the retina and sent to the brain. The number of possible connections to the brain exceeds $10^{80}$ (a "1" followed by 80 zeros), about as many atoms as there are in the entire universe. The entire retina contains not only these 15,000 pixels, but 126 million of them, the signals of which are then reduced in the retina to about one million. Nevertheless, an accidental origin of the order of these nerve fibres can certainly be considered impossible (1).

When we consider spatial perception as well, the matter becomes even more complicated. To enable the brain to produce a spatial depth impression, both eyes must see the same image. The images seen by the two eyes contain systematic differences due to the spatial relations. The brain calculates the distance based on the differences between the pixel signals from the two different eyes. This is done individually for each pixel. Spatial depth, i.e., binocular vision, is only possible if the nerve fibres end exactly at the right location in the brain.

Without doubt, a most ingenious intelligence was required to write into the DNA the program for making such a highly-ordered system (2).

# References

1. David E. Stoltzmann. "The Specified Complexity of Retinal Imagery." *CRSQ* 43/1 (Juni 2006): 4–12.
2. Wolf-Ekkehard Lönnig. Auge widerlegt Zufalls-Evolution, 2. Auflage, Naturwissenschaftlicher Verlag Köln, 1989, http://www.weloennig.de/AuIEnt.html.

# 90

# THE INVERSE RETINA

The light-sensitive cells in the human eye are located beneath two layers of nerve cells. It appeared that the light would therefore be weakened by the nerve cells. An intelligent creator, it was thought, would have used a more efficient construction. However, it then turned out that the so-called Muellerian cells, hitherto considered to have a supporting function only, also function as highly efficient light conductors that transmit the light between the nerve cells to the light-sensitive cells in the retina. Since the light-sensitive cells lie directly above the blood vessels, they are cooled better and can be provided with energy in a more efficient manner.

In the human eye, the surface layers of the retina contain nerve cells, beneath which lie the light-sensitive rods and cones. Since the nerve cells lie above the rods and cones, they would under normal circumstances appear to weaken the light, thus interfering with vision. For this reason, evolutionary researchers claimed that this arrangement could not possibly have been conceived by an intelligent creator.

Recent research results obtained at the Paul Flechsig Brain Research Institute at the University of Leipzig have now shown that light is not scattered or lost at all in the human eye. So-called Muellerian cells conduct the light from the front surface of the retina to the light-sensitive cells embedded in the back of the retina, whereby their function is similar to that of a fibre optic cable. The light is thus transmitted without losing intensity between the nerve cells through to the light-sensitive cells. The conical shape of the Muellerian cells collects light instead of scattering it. What this means is that our vision is optimized by just this specific arrangement of nerve cells, Muellerian cells, rods and cones (1).

It therefore makes sense that the light-sensitive cells lie beneath the others because they require the most energy, of which they are ensured an optimum supply by their location right above the blood vessels. Another factor is that the blood vessels also cool the light-sensitive cells, preventing damage to the retina that could otherwise be caused by infrared radiation (2). In squids the cells are arranged in the opposite order because these animals live in cool water. In this case, it does make sense to position the light-sensitive cells in the top layer, since the eyeball is cooled by the water (3).

To sum up, the different designs of human and squid eyes ensure optimum vision for both creatures and provide clear evidence of an intelligent and consummate creator of these two systems.

# References

1. Kristian Franze, et al. Müller cells are living optical fibres in the vertebrate retina, herausgegeben von Luke Lee, University of California, Berkeley, CA, and vom wissenschaftlichen Beirat am 27. März 2007 angenommen, http://www.pnas.org/cgi/content/short/104/20/8287.
2. Sylvia Baker. "Seeing and believing." *Genesis Agendum* (2004): 4.
3. Willian A. Dembski and J.M. Kushiner. *Signs of Intelligence*. (Bazos Press, 2002): 216.

# 91

# THE DEGENERATION OF HUMAN LANGUAGE

Investigations of ancient languages show that they were more complex in earlier times and became simpler over time. The following holds for ancient Latin, Greek, Hebrew, Chinese, and Native American languages. For as far back as we can see, early human languages were able to communicate more information with fewer words than is the case with modern languages. Also, more precise formulations were possible with these languages. This contradicts the evolutionary idea of development from simple beginnings to greater complexity.

It has turned out that the idea of development of human language from primitive beginnings to a more sophisticated state is not upheld by the evidence. The languages of so-called aboriginal natives are not the least bit primitive. They are highly complex, in most cases much more complicated than our European languages.

Research into aboriginal languages has made it clear that there is no connection between the cultural level of a society and the structure of its language. This means a tribal society can live under the simplest conditions imaginable and still use an extremely complex language.

The complex structures of Old Sumerian, Old Akkadian and Old Egyptian contrast impressively with the, in some cases, comparatively very simple morphological structures of the modern languages spoken in Europe today. Whereas Akkadian, for instance, had thousands of synthetic verbal forms, modern German has a comparatively small inventory of forms (1). The term "synthetic verbal forms" designates linguistic forms comprising a single word and requiring no further auxiliary verbs (e.g., have, be, want to, may) to complete or complement their meaning.

## Egyptian, Akkadian, Hebrew and Greek

Roger Liebi carried out a study of very old languages of which we have written documentation covering long periods of history. The following were among the languages investigated: Egyptian (over 4,000 years), Akkadian (2,600 years), Hebrew (3,500 years) and Greek (3,500 years). Liebi concluded, "Everywhere you look in the history of languages, the development you always see is characterized by degradation, reduction and simplification, i.e., in particular in morphology and the related field of phonology. The history of language can essentially be said to be characterized by devolution (backwards or downwards development) in the areas of morphology and phonology" (2).

Liebi sees the reason for this in the laziness of speakers. This laziness results in the casting off of phonological elements to the point of elimination of morphological structures.

## The Tower of Babel

The biblical report tells us that all of the people living after the great flood spoke a single language. As they began to spread abroad, they said to one another, "Come, let us build ourselves a city, and a tower with its top in the heavens, and let us make a name for ourselves, lest we be scattered abroad upon the face of the whole Earth." As a punishment for this hubris, God then confused their language and scattered them abroad from there over the face of all the Earth, no longer understanding one another (3).

An interesting aspect of this account is that all of the cultures that arose at that time all over the Earth carried out astronomical calculations and erected in some cases gigantic monuments. Each culture possessed part of the totality of human information. It can be assumed that human languages have been degenerating ever since.

It would appear to approach nearer to the actual situation to seek the origins of human language in a single, highly complex language (which may have been vastly superior to our modern languages) than to derive them from the sounds made by animals (4).

## References

1. Roger Liebi. The Mensch, ein sprechender Affe?, Schwengeler Verlag, 1991, 48.
2. Roger Liebi, 52.
3. *The Bible.* Gen. 11:1–9.
4. Roger Liebi. Herkunft und Entwicklung der Sprachen, Hänssler, 2007, 272–276.

## 92

# HUMAN CONSCIOUSNESS

So-called near-death experiences suggest that human consciousness can exist separately from the body. It is true that even consistent first-hand reports by people who were clinically dead for brief periods cannot provide absolute certainty, because human consciousness is a phenomenon that cannot be definitively defined, either in medical or philosophical terms. But each person can make his or her own decision to answer the question: Is your consciousness the product of an essentially dead mechanism, or is it a part of your original "self" that exists independently of your physical body?

Consciousness is understood to be the ability to have thoughts, emotions, perceptions and memories and to be aware of them, experience them, be conscious of them. The phenomenon of consciousness is often considered one of the biggest unsolved problems of philosophy and natural science. There is currently no precise and generally recognized definition of consciousness.

The actual puzzle presented by consciousness can be expressed by the question of how it can be possible in principle that the specific arrangement of molecules and dynamics in the active brain results in the actual awareness that is consciousness. The question is not so much how our brain processes the signals from the nerve cells and how we react to this, but rather where does this awareness end? Who or what is, in the final analysis, the recipient of the content of the experiences presented by the brain? Who am "I" that I experience and live all of this?

Does a human being have a supernatural spirit and does this spirit continue to exist in some form of consciousness when physical brain activity stops?

The puzzling nature of the phenomenon of consciousness has two different aspects:

For one thing, states of consciousness have an experiential content, and it is not clear how the brain can produce experience, nor is it clear who or what it is, in the final analysis, that takes in this experience and actually experiences it. This is the so-called qualia problem (1).

The other aspect is that thoughts not only refer to objects, but to empirical contents as well. The imagination imagines something, judgement recognizes or rejects something, love loves, hate hates, and desire desires something. The thought that there is still some milk in the refrigerator refers to the objects "refrigerator" and "milk" – and to the factual content that there is still milk in the refrigerator. It is a complete mystery how the brain can produce thoughts with such properties – and who or what finally takes in this content and realizes it in the truest sense. This is the so-called intentionality problem.

"I can explain my body and my brain, but there is more. I cannot explain my own existence," said Australian brain researcher and Nobel Prize winner John C. Eccles. Eccles sought for answers to, among other things, the question of how nerve cells conduct stimuli and make major contributions to understanding the processes at work in the human brain.

Eccles also addressed the problem of consciousness in philosophical terms. He himself held the opinion that only humans possess ego consciousness, a feature inherent in a human being from the moment of conception that develops in relationship to the outer world during the first years of life. He rejected strict materialism (the assumption that consciousness could be explained solely on the basis of physical and chemical processes) and compared the brain to a computer and the self to its programmer. He considered that this self (spirit, soul) was supernatural and used the brain as its instrument. For this reason, he thought, we could hope that the self might continue to exist after death (3).

Eccles was known in particular for the 1977 publication of *The Self and Its Brain*, which was written together with Karl Popper.

## References

1. David Chalmers. *The Conscious Mind*. Oxford University Press, 1996.
2. John Searle. *Intentionality - An Essay in the Philosophy of Mind*. Cambridge University Press, 1983.
3. John C. Eccles, *factum* 5/2001, 17.

# 93

# HUMAN CREATIVITY

Creativity creates something new. Not only artists are creative: automotive engineers, road building engineers, programmers, housewives, and students all find solutions to complex problems and create things that did not exist before. Our ability to create new things and to figure out the secrets of the universe and the material world could derive from the fact that we are creatures made in the image of the God who created it all.

Human creativity is practically unlimited when it comes to researching and creating complex systems. The Bible says that we were made second only to God (1). So were we created in God's own image, in the image of the creator of the entire universe (2)? Or are we ourselves the highest authority on Earth? Did we create the idea of God? Or are we all thoughts in His mind?

According to evolutionary theory, one might assume that we are superior to all other creatures on Earth. Richard Dawkins was thinking along similar lines when he wrote that "any creative intelligent sufficiently complex to form and create something can only come to be as the final product of a long process of gradual evolution" (3).

In contrast to this, the Bible says that God, the creator of all human abilities "Is, who He is" and has been for all eternity. Yahweh, the Hebrew name of God, is translated as, "I am that I am."

If we imagine God to be an eternal spirit and an eternal founding principle, He cannot be said to have developed over time. When the proponents of evolution state that life will necessarily come to be on any planet that possesses the necessary preconditions, they are also stating a belief in a founding principle of life that "is what it is" and has always been and inevitably creates life. In the final analysis, they believe in the same cause as a person who believes in God, namely a cause of all things that has existed for all eternity.

## Cause and effect

Everything that arises from a previous something must have been contained in that previous something in some form.

A cause may have many different effects, but none of the effects can be quantitatively larger or qualitatively more enhanced than the cause. This is made clear by the law of conservation of energy, the first law of thermodynamics.

Since we humans possess consciousness, it makes sense to conclude that the cause of our existence would also possess consciousness. Energy comes only from energy, life comes only from life, and consciousness comes only from consciousness. This is certainly plausible. Or is it conceivable that we, the final product of a long development, should be the first to think about our own existence and be able to attain to consciousness of ourselves? There are also some evolutionists who hold the opinion that we are not the most highly-developed creatures in the universe. The aliens in whom many of them believe are supposed to be superior to humans on Earth, comparable to the supernatural beings of the religions.

Aspects that must not be forgotten in relation to the developments we are witnessing today:

- a) How is it that living beings can adapt to their surroundings? They can do this because they already have pre-programmed mechanisms that facilitate such adaptation.

b) Why have human technologies made progress? Because humans already have a creative brain, a creative mind.

## The human brain

The different opinions entertained over the years in the history of brain research make it clear how little we still know concerning the function of our brain and our cognitive abilities.

In the middle of the nineteenth century, physician Rudolf L.K. Virchow discovered the so-called glial cells. His assumption that these cells served to support and hold other structures was the basis for the name he gave them. Glia is derived from the Greek for "glue." Human nervous tissue contains a much higher proportion of glial cells than a doe's animal nervous tissue. Glial cells are smaller than nerve cells and make up about fifty percent of the brain. The human brain contains about ten to fifty times more glial cells than neurones.

Until recently it was thought that these cells form a supportive framework for the nerve cells and at the same time provide the mutual electrical insulation required by the nerve cells. More recent insights have shown that glial cells play an important role in mass and fluid transport as well as in the maintenance of cerebral homeostasis.* They also contribute to information processing, storage, and conduction.

The discovery of these functions a few years ago marked the real beginning of research on the human brain.

\* Homeostasis (balancing capacity) refers to the constant effort exerted by an organism to interadapt various different physiological functions and maintain a constantly-balanced status for as long as possible. In this respect, reference is made in particular to one of the smallest regions in the brain, the so-called hypothalamus, an overriding signal-switching centre located at the base of the brain that is also an important integrative organ for regulation of the body's internal milieu.

## References

1. *The Bible*, Psalms 8:6.
2. *The Bible*, Gen. 1:27.
3. Richard Dawkins. *The God Delusion*. Bantam Press, 2006.

# 94

# CONSCIENCE AND ETHICS

Conscience and ethics are hardly things that would have evolved in a merciless fight for survival that has been going on for millions of years. Conscience does not increase one's chances of survival. Pure instinct deprived of conscience would probably result in most cases in the elimination of the enemy race. Conscience, on the other hand, keeps one from acting on purely unscrupulous or selfish motives.

The world of evolution (as formulated by evolutionists themselves) is a world of pure accident devoid of essential being. Life and death, to be or not to be – everything is of equal value because everything happens purely by accident, without any plan or objective. It would most certainly be difficult to anchor ethical obligations in a meaningless and erratic world without the solid foundations.

Evolutionary ethics as they generally are formulated (and which are by definition without meaning or objective) still strive according to some famous philosophers towards an overall goal. Friedrich Nietzsche formulates this as development into a "Superman" and Teilhard de Chardin (with a pseudo-Christian touch) speaks of the "Omega point." From the point of view of evolutionary theory, assisted suicide, abortions and some genetic manipulations (as instruments of further evolutionary development) could be assessed quite positively. The conscience, on the other hand, is generally opposed to such practices (1).

If the human race would really be the product of a pitiless evolutionary fight for survival there would be no reason to hinder evolutionary progress with ethics and morals. The survival of the fittest cannot be the basis for our ethics. How could ethical questions even occur to us if it were really the case that the egoistic urge to survive had been the ruling principle of animal behaviour for many millions of years? How could it be that egoistic creatures would suddenly start thinking of the well-being of others?

In the animal world (2) we see individuals cooperating to hunt, fend off enemies and care for sick family members. Reciprocal altruism (feeding and delousing each other, symbiosis involving two different species) is frequently observed among animals. Human ethics, on the other hand, is concerned with thinking about the morally correct thing to do regardless of one's own advantage. Human ethics comprise a reflection on what is really the right thing to do and not what would be most advantageous to me or my family in this moment (utilitarianism).

Most European philosophers are not consistent in ethical matters. Their scientific background is the theory of evolution, but their ethics and principles of practical action are still based (either consciously or unconsciously) on the Bible. A quotation from English naturalist Thomas Henry Huxley from the nineteenth century is interesting in this connection, "I must admit that I was confused when I set out to establish foundations for moral behaviour in our chaotic times without using the Bible" (3).

Apparently it is not possible to base ethics worthy of this name on the theory of evolution. The attempts that have been made in this direction read like a list of the darkest periods of human history: Hitler and National Socialism; Marx, Stalin and Communism. The worst crimes in the history of humanity have been committed under the erroneous belief of the theory of evolution.

## References

1. Marcel Wildi. Evolution und Schöpfung und die jeweiligen Konsequenzen für die Ethik, Seminararbeit, STH Basel, 1992.
2. Frans de Waal. Der gute Affe, dtv, München, 1996.
3. Octobible-Führer der Expo Tabernacle, Lausanne, 1992, 15.

## 95

## LOVE, JOY, SUFFERING AND SORROW

It is hard to reconcile the existence of the phenomenon of love with the theory of evolution. Aside from sexual love, it is an indescribable, purely spiritual component that contradicts the naturalistic world principle. Could life really have arisen from inanimate, insentient matter? If that is the case, love, joy, suffering, and sorrow are nothing more than immensely complicated naturalistic mechanisms that would be more of a hindrance than a help in the pitiless struggle for survival. Could it have been that on the beginning of life not pure chance could have ruled but in fact the love of an intelligent designer?

In the narrower sense, love designates the deepest sense of attachment that a person can have for another one. Love is a feeling, or perhaps even more an inner attitude, of positive, sincere and deep devotion to a person that goes beyond the mere purpose or usefulness of an interpersonal relationship and is normally expressed as active affection for the loved one.

### Ancient Greek differentiates between three kinds of love

- Eros is sensual, erotic love, the desire of the loved one, the wish to be loved, and passion.
- Philia is the term for love of family and friends, mutual love, mutual recognition and mutual understanding.
- Agape is selfless, helping love, the love of a father or mother for their child, brotherly love, love of one's

enemy, love that looks to the well-being of the beloved. It is also known as the love of God (1).

In evolution, relationships are characterized mainly by egoisms, whereas agape and philia look first to the well-being of the other person. Even love based on Eros is controversial in evolutionist circles. Quite a number of proponents of the theory of evolution make reference to the many unisexual life forms that require no Eros-based love at all. But why do these forms exist? To ensure the survival of the fittest individuals? Not a very romantic notion ...

## Conclusion

The age of the Earth and the universe can be investigated using scientific methods, as can the descent of species and the structure of our ecosystem. Such investigation must seriously call into question the models offered by the evolution, primordial soup, and big bang theories. At any rate, no one will be able to prove by scientific methods that a loving and caring creator embraces all of life and loves you personally from the bottom of His heart, a creator who shares your joys, sufferings and sorrows and who would like to take you back into His bosom when your physical heart has ceased to beat.

Imagine a young man in love who would like to prove to his girlfriend that he loves her. Will he try to do this by means of logical arguments and scientific methods? Would a love with a purely logical essence be worthy of the name love at all? Nor could it be God's plan to give us rational answers to all our questions. For love is not grasped with the rational mind, but rather with the heart. God is love. If you believe that, you already know a great deal more than science (no matter how sophisticated) will ever assess.

## Reference

1. *The Bible*. 1 Cor. 13.

# FINAL DECLARATION

We have written these ninety-five theses to the best of our knowledge based on the broadest possible range of source material. Nonetheless, our knowledge is far from complete and certainly this and subsequent versions of the ninety-five theses may still contain errors.

We ask you to help us in our efforts to correct such errors. We are asking you please to send us feedback and proposed changes in accordance with state-of-the-art. You will find our contact information, along with the latest version of the theses, under www.0095.info

## THE AUTHORS:

Dr. iur. Dieter Aebi, Dr. med. Markus Bourquin, Prof. a. D. Dr.Ing. Werner Gitt, Roland Schwab, Dipl.Ing. Hansruedi Stutz, lic. theol. Marcel Wildi.

# EPILOGUE

The work of both evolutionists and creationists is based on exactly the same scientific data. It is not a matter of scientific data but rather of one's world view which influences the interpretation and extrapolation of said data – consciously or unconsciously – whether one accepts or rejects the models presented by the theories of evolution, the primordial soup and the big bang.

The belief that chemical and physical laws could suffice to bring forth the complexity and variety of life and the immeasurable cosmos is not supported by the data the natural sciences have provided. Many questions about origins have to be answered in natural science by simply saying we don't know. There is more honesty to this response than to keep insisting that unproven hypotheses are proven facts.

The authors are well aware that there are many researchers doing basic research in the natural sciences, making an extraordinary personal effort to widen the horizons of our knowledge. These efforts not only produce answers, they also produce many new and unexpected questions. The number of unsolved questions is increasing more rapidly than the number of answered questions.

An unfortunate aspect of the public debate on evolution is that while unclarified details of the theory are discussed, evolution in the sense of progressive development is not called into question per se. Persons who questions the above-mentioned theories are threatened with exclusion from scientific and educational positions. Within such a framework, proponents of the theory of evolution often, albeit unconsciously, assume a totalitarian, dogmatic and ideological stance in their argumentation.

The authors of the ninety-five theses long for a society in which every person would be given the opportunity to choose and defend his or her world view independently as long as this does not restrict others' freedom. They consider it legitimate to reflect upon an alternative world view regarding the origins of life, free of evolutionary dogmas and societal pressures.

The ninety-five theses presented here reveal the basic insufficiency of the claims of the evolutionary hypothesis. The orientation of the author's approach to this subject is the world view of the Bible, in which it is written, "Ever since the creation of the world his invisible nature, namely, his eternal power and deity, has been clearly perceived in the things that have been made." (1)

# Reference

1. *The Bible*, Rom. 1:20.

www.ingramcontent.com/pod-product-compliance
Lightning Source LLC
Chambersburg PA
CBHW020731180526
45163CB00001B/192